Ascensions, Lives & Revelations of Old Testament Figures

Legendary Narratives of Adam, Jacob, Joseph, the Prophets, and More

A Modern Translation

Adapted for the Contemporary Reader

Various Ancient Writers

Translated by Tim Zengerink

© **Copyright 2025**
All rights reserved.

It is not legal to reproduce, duplicate, or transmit any part of this document in either electronic means or in printed format. Recording of this publication is strictly prohibited and any storage of this document is not allowed unless with written permission from the publisher except for the use of brief quotations in a book review.

This book contains works of fiction. Any resemblance to persons living or dead, or places, events, or locations is purely coincidental.

Table of Contents

Preface - Message to the Reader 1
Introduction 4
Life of Adam and Eve 9
    Introduction 9
    Life of Adam and Eve 15
Ladder of Jacob 28
    Introduction 28
    Ladder of Jacob 33
Joseph and Asenath 39
    Introduction to Joseph and Asenath 39
    Chapter One 39
    Chapter Two 41
    Chapter Three 42
    Chapter Four 43
    Chapter Five 44
    Chapter Six 45
    Chapter Seven 46
    Chapter Eight 47
    Chapter Nine 48
    Chapter Ten 49
    Chapter Eleven 50
    Chapter Twelve 51
    Chapter Thirteen 52
    Chapter Fourteen 53

Chapter Fifteen 55
Chapter Sixteen 56
Chapter Seventeen 58
Chapter Eighteen 59
Chapter Nineteen 60
Chapter Twenty 60
Chapter Twenty-one 61
Chapter Twenty-two 62
Chapter Twenty three 63
Chapter Twenty-four 65
Chapter Twenty-five 67
Chapter Twenty-six 68
Chapter Twenty-seven 69
Chapter Twenty-eight 70
Chapter Twenty-nine 72

Revelation of Elijah 74

   Introduction 74
   Revelation of Elijah 79

Revelation of Zephaniah 93

   Introduction 93
   Revelation of Zephaniah 98

Thank You for Reading 109

Preface - Message to the Reader

What If You Could Help Rebuild the Greatest Library in Human History?

Thousands of years ago, the Library of Alexandria stood as the crown jewel of human achievement — a sanctuary where the collected wisdom of every known civilization was gathered, preserved, and shared freely.

And then, it was lost.

Through fire, conquest, and the slow erosion of time, humanity lost not just books — but ideas, dreams, discoveries, and stories that could have changed the world forever.

Today, the Library of Alexandria lives again — and you are invited to be a part of its restoration.

Our mission is simple yet profound:

To rebuild the greatest library the world has ever known, and to translate all timeless works into every language and dialect, so that no seeker of knowledge is ever left behind again.

By joining our movement to rebuild the modern Library of Alexandria, you become part of an unprecedented mission:

- **Unlimited Access to the Greatest Audiobooks & eBooks Ever Written:**

    Instantly explore thousands of legendary works—Plato, Shakespeare, Jane Austen, Leo Tolstoy, and countless more. All instantly available to read or listen, placing a complete literary universe at your fingertips.

- **Beautiful Paperback & Deluxe Editions at Printing Cost**

    Own any title as an elegant paperback, deluxe hardcover, or stunning collectible boxset—offered to you at true printing cost, delivered straight to your door. Build your personal Library of Alexandria, crafted for beauty, built for durability, and worthy of proud display.

- **Fresh Translations for Modern Readers—in Every Language & Dialect**

    Enjoy timeless masterpieces reimagined in clear, contemporary language—no more outdated phrases or obscure references. Alongside the original versions, we're tirelessly translating these classics into every language and dialect imaginable, ensuring accessibility and understanding across cultures and generations.

- **Join a Global Renaissance of Literature & Knowledge**

    You directly support expanding our library, publishing deluxe editions at true cost, translating works into all global languages, and bringing humanity's greatest stories to people everywhere. By joining today, you're not just preserving a legacy of masterpieces; you set in motion a powerful wave of literary accessibility.

Become a Torchbearer of Knowledge.

Join us for free now at **LibraryofAlexandria.com**

Together, we will ensure that the light of human wisdom never fades again.

With gratitude and a shared love of knowledge,
The Modern Library of Alexandria Team

Visit:

www.libraryofalexandria.com

Or scan the code below:

Introduction

Beyond the Canon:
Rediscovering Ancient Legends and Sacred Visions

The Hebrew Bible offers a profound spiritual narrative—rich in creation myths, covenantal promises, moral law, prophetic rebuke, and messianic hope. But what if the stories of Genesis, the Patriarchs, and the Prophets were only the beginning? What if the canonical texts we know today were once accompanied by vibrant expansions—tales passed down through oral tradition, elaborated in temple schools, and cherished by ancient readers hungry for more? These expanded accounts, now called pseudepigrapha or apocryphal writings, preserve a remarkable tapestry of sacred imagination. In Ascensions, Lives & Revelations of Old Testament Figures, we enter this parallel world of Jewish narrative tradition, a universe of dramatic elaborations, mystical visions, and legendary epics.

While these texts were not included in the official canon, they were revered by many ancient communities. Their popularity persisted for centuries across Jewish, Christian, and even Islamic circles. They do not rewrite scripture so much as magnify it—pulling back the curtain to reveal what lies beyond the sparse narrative lines of the Bible. In them, Adam and Eve weep over paradise lost, Jacob dreams of heaven in vivid symbolic detail, Joseph finds love and struggle with Asenath, and prophets like Elijah and Zephaniah glimpse realms of divine judgment and redemption.

This collection is not merely a gathering of forgotten curiosities. It is a profound testament to humanity's yearning to understand the divine, to see the face of God, and to live lives of moral clarity and cosmic purpose. It provides an indispensable bridge between the Hebrew Bible and later Jewish mysticism, early Christian visions, and Islamic eschatological traditions.

Mythic Origins and Sacred Destinies

Among the most captivating texts in this volume is the Life of Adam and Eve, known in some versions as the Apocalypse of Moses. This narrative picks up where Genesis ends, chronicling the sorrowful exile of the first humans from Eden, their repentance, and their eventual deaths. It paints Adam not only as a father of humanity but as a spiritual archetype—struggling with regret, longing for mercy, and awaiting a glimpse of future redemption. Eve too is given a voice, offering her own account of the fall and her hope for salvation. These texts explore the themes of divine justice, human frailty, and the promise of restoration.

Equally compelling is the Ladder of Jacob, which expands the brief biblical account of Jacob's dream into a sweeping apocalyptic vision. Angels descend and ascend not just in a symbolic gesture, but as participants in a cosmic revelation. Jacob is given insight into the future of Israel, the nature of divine warfare, and the spiritual realms that lie beyond mortal comprehension. In this ladder, readers ascend alongside Jacob into a mystical theology that anticipates both Kabbalistic thought and Christian heavenly ascent literature.

The Joseph and Asenath narrative is another gem—an elegant novella that blends romance, conversion, and divine destiny. It imagines how Joseph, the noble son of Jacob, married Asenath, an Egyptian priest's daughter who undergoes a profound spiritual transformation. Their love story becomes a metaphor for reconciliation between Israel and the nations, for purity, and for the triumph of righteousness. This text resonated deeply with early Christian readers, who saw in Asenath a symbol of the Gentile Church, purified and wed to the Messiah.

The Lives of the Prophets presents a series of biographical sketches of figures like Isaiah, Jeremiah, Ezekiel, and Daniel. These are not dry entries, but living portraits—filled with signs, wonders, visions, and martyrdoms. Here, prophets do not merely proclaim God's word; they embody it. These brief accounts preserve ancient traditions about where the prophets lived, how they died, and what miracles marked their ministries. Complementary texts like the Revelation of Elijah and Revelation of Zephaniah carry readers into prophetic dreamscapes where angels interpret history, demonic forces are unveiled, and divine mysteries are made known.

Collectively, these writings form a body of literature that might be called a "midrashic apocalypse"—combining deep devotion to the Hebrew scriptures with imaginative expansions and mystical insight. They preserve the memory of spiritual heroes while offering bold new visions of their journeys, trials, and revelations.

The Role of These Texts in Religious and Literary History

Although labeled "apocryphal" by later authorities, many of these texts were widely read and copied in Jewish and early Christian communities. Manuscripts appear in Greek, Latin, Syriac, Ethiopic, Armenian, and Slavonic, indicating their broad circulation and influence. Church fathers such as Origen and Ambrose referenced them, and they shaped medieval legends, liturgy, and mystical traditions. They also appear in Islamic hagiography, especially concerning Adam, Moses, and Elijah.

Their popularity stemmed from their ability to answer lingering questions: What happened after Eden? How did patriarchs die? What secrets did the prophets know? Why did God choose certain individuals and not others? And most significantly: What lies beyond this life? These texts filled narrative gaps, resolved theological tensions, and provided spiritual comfort in the face of empire, persecution, and uncertainty.

For modern readers, these stories are not simply historical artifacts. They are invitations to spiritual reflection. They challenge us to read between the lines of scripture, to appreciate the power of sacred imagination, and to reengage with foundational moral themes. The texts offer timeless warnings against pride, idolatry, and injustice—while holding out hope for restoration, forgiveness, and divine intimacy.

They also illuminate the porous boundaries between Jewish, Christian, and Gnostic traditions. The visionary accounts of Elijah and Enoch, the heavenly travels of Isaiah and Zephaniah, and the teachings of angelic intermediaries echo across religious boundaries. These shared traditions remind us that the spiritual heritage of the ancient world was more fluid and interconnected than later doctrinal boundaries might suggest.

This collection therefore serves multiple purposes. It is a spiritual treasury, a historical witness, a literary exploration, and a theological catalyst. Whether you are a believer, a scholar, or simply a curious seeker, these texts invite you into a deeper engagement with the characters you thought you knew—and with the God who walked beside them.

May the lives, ascensions, and revelations gathered here draw you closer to the heart of the biblical world—not only in its sacred stories,

Translated by Tim Zengerink

but in its enduring hunger for truth, justice, mercy, and the divine presence that surpasses all understanding.

Life of Adam and Eve

Introduction

The Life of Adam and Eve is one of the most fascinating and influential texts from early Jewish and Christian literature, renowned for its profound exploration of humanity's origins, morality, suffering, redemption, and the complex relationship between humanity and the divine. This ancient work, known primarily through several textual traditions—including Greek, Latin, Armenian, Georgian, Slavonic, and other manuscripts—expands richly upon the Genesis narrative, vividly portraying the experiences, emotions, struggles, and reflections of Adam and Eve after their expulsion from Eden. By elaborating on their lives beyond the brief biblical account, this text provides a deeper and more intricate exploration of fundamental human themes such as guilt, repentance, mortality, family, spiritual longing, and divine mercy.

Though not part of the canonical scriptures, the Life of Adam and Eve significantly impacted Jewish, Christian, and even Islamic religious thought, influencing theology, popular devotion, liturgy, and visual arts throughout the centuries. Its vivid storytelling, emotional resonance, and profound spiritual reflections captured the imagination of countless generations, making it one of the most widely disseminated and influential extra-canonical texts in antiquity and beyond. In its varied textual traditions, the Life of Adam and Eve offers readers remarkable insights into ancient interpretations of humanity's origin, the nature of sin, the dynamics of divine judgment and forgiveness, and the complexities of human existence within a fallen world.

This introduction aims to provide readers with a comprehensive historical, literary, and theological framework necessary for deeply engaging with the Life of Adam and Eve. By exploring its origins and textual history, analyzing its rich narrative content and theological themes, and examining its enduring legacy and cultural influence, readers will be well-prepared for an enriching and insightful reading experience. Despite its varied manuscript traditions and complex textual history, the Life of Adam and Eve remains a profoundly unified and spiritually evocative work, offering valuable lessons and reflections that resonate deeply with contemporary readers seeking to understand humanity's perennial struggles with sin, mortality, redemption, and spiritual yearning.

Historical Context and Textual Traditions

Understanding the historical origins and textual traditions of the Life of Adam and Eve is essential to fully appreciating its richness and complexity. While precise dating remains uncertain, most scholars agree that the text originated between the first century BCE and the third century CE, a period marked by tremendous religious creativity, cultural interaction, and theological reflection within Judaism and emerging Christianity. Composed originally in a Jewish context, the text reflects theological concerns, ethical teachings, and spiritual reflections characteristic of late Second Temple Judaism, resonating with other pseudepigraphal and apocryphal writings that expanded upon canonical biblical narratives.

Although the original language and exact provenance of the text remain uncertain, scholars commonly propose Hebrew or Aramaic origins, subsequently translated into Greek and other languages. The surviving manuscripts reveal a complex textual history, preserved across multiple languages, cultures, and religious traditions, each textual tradition exhibiting unique emphases, variations, and narrative expansions. Among the various versions, the Greek text—often referred to as the Apocalypse of Moses—and the Latin tradition, known as the Vita Adae et Evae, are most influential and widely studied. These diverse manuscript traditions testify to the text's broad popularity, adaptability, and profound influence across diverse linguistic and cultural contexts.

The text's widespread dissemination and translation into multiple languages further indicate its enduring spiritual appeal, theological significance, and emotional resonance among early Jewish and Christian communities. The Life of Adam and Eve served as an important devotional and instructional resource, providing communities with profound reflections on fundamental theological and existential questions, such as the origins of human suffering, the nature of sin, the possibility of repentance, and the promise of divine redemption. The complexity of its textual history thus underscores both the text's universal spiritual relevance and its specific cultural adaptations, demonstrating its lasting significance within religious, theological, and literary traditions across centuries and cultures.

Narrative Content, Themes, and Theological Reflections

At the heart of the Life of Adam and Eve lies a rich narrative expansion of the biblical Genesis story, vividly exploring Adam and Eve's lives following their expulsion from Eden. While the canonical Genesis account briefly narrates humanity's fall, expulsion, and initial hardships, the Life of Adam and Eve delves deeply into their emotional experiences, moral reflections, interpersonal struggles, and ongoing spiritual quest for forgiveness, reconciliation, and divine mercy. By elaborating upon these themes, the text provides readers with profound insights into universal human experiences such as guilt, suffering, family relationships, spiritual aspiration, repentance, and redemption.

The text's narrative structure follows Adam and Eve as they grapple deeply with the painful consequences of their disobedience, vividly portraying their remorse, shame, and longing for restoration. Readers encounter intense scenes depicting Adam and Eve's profound emotional and spiritual struggles—grief over paradise lost, conflicts within their family, attempts at reconciliation, and fervent prayers for divine mercy. These narrative episodes emphasize not merely the physical hardships of life outside Eden but also the intense emotional, moral, and spiritual dimensions of their new existence, profoundly humanizing the biblical characters and making their experiences deeply relatable to readers.

Central theological themes within the Life of Adam and Eve include reflections on the nature of sin, divine judgment, human responsibility, and the possibility of redemption. The text explores humanity's original disobedience not merely as historical transgression but as ongoing spiritual struggle, highlighting sin's complex effects on human relationships, self-awareness, and spiritual longing. Adam and Eve's continual repentance and prayers for forgiveness underscore the dynamic relationship between divine justice and divine mercy, offering profound theological insights into the character of God as both just judge and merciful redeemer. By vividly portraying divine encounters, angelic visitations, and moments of revelation, the text emphasizes divine compassion, justice, and enduring faithfulness despite human failings.

Moreover, the Life of Adam and Eve deeply explores the relational dynamics within the first human family, particularly emphasizing Adam and Eve's relationship, their interactions with their children, and the complexities arising from sin's lingering consequences. Narratives highlighting familial conflicts, reconciliation efforts, and intergenerational dynamics provide profound reflections on human relationships, moral responsibilities, and the enduring impact of parental choices on subsequent generations. By emphasizing these relational dimensions, the text invites readers to reflect deeply on universal human experiences of conflict, reconciliation, moral legacy, and communal identity.

Legacy, Influence, and Contemporary Significance

The historical and cultural legacy of the Life of Adam and Eve is extensive, reflecting its widespread influence across diverse religious traditions, theological reflections, liturgical practices, and artistic expressions. Throughout Jewish, Christian, and even Islamic religious history, the text profoundly impacted theological discussions concerning human origins, sin, suffering, and redemption. Its rich narrative expansions, vivid portrayals, and profound theological insights provided invaluable resources for theological reflection, spiritual meditation, ethical instruction, and devotional practice across numerous religious communities.

In Christian traditions, the text significantly influenced theological reflections on original sin, human mortality, and divine redemption, shaping early Christian doctrines and devotional practices. Its vivid portrayals of Adam and Eve's emotional struggles, penitential prayers, and relational dynamics profoundly resonated with medieval Christian spirituality, influencing popular devotions, liturgical commemorations, and artistic representations of biblical narratives. Similarly, Jewish mystical and ethical traditions drew extensively from the text's reflections on humanity's spiritual struggles, ethical responsibilities, and relational complexities, incorporating its insights into ethical teachings, mystical symbolism, and devotional practices.

Moreover, the Life of Adam and Eve profoundly influenced visual arts, literature, and popular culture, inspiring numerous artistic representations, literary adaptations, and cultural references throughout history. Its vivid narrative scenes, emotional depth, and theological resonance provided powerful source material for artistic exploration, theological reflection, and cultural engagement, testifying to the text's enduring spiritual relevance and cultural influence.

Today, contemporary scholarship continues to explore the text's historical significance, textual variations, theological insights, and cultural impact, highlighting its profound relevance for understanding ancient religious thought, theological reflection, and spiritual experience. For modern readers, the text offers valuable insights into timeless human struggles with guilt, suffering, spiritual aspiration, relational complexities, and divine-human interaction, reflecting deeply universal human experiences and spiritual questions. Its exploration of sin's complexities, redemption's possibilities, and relational dynamics remains profoundly resonant for contemporary spiritual reflection, ethical instruction, and cultural dialogue.

In conclusion, the Life of Adam and Eve stands as a profoundly significant and spiritually evocative text within early Jewish and Christian literature, offering modern readers invaluable insights into humanity's perennial spiritual struggles, relational complexities, and theological reflections. By engaging deeply with this remarkable text, contemporary readers are invited into profound dialogue with ancient religious traditions, theological insights, and universal human experiences, enriching modern understandings of spirituality, morality, redemption, and divine mercy. The enduring spiritual relevance, theological depth, and emotional resonance of the Life of Adam and Eve ensure its continuing significance within contemporary religious, cultural, and spiritual exploration.

Life of Adam and Eve

After Adam and Eve were sent away from paradise, they built a small shelter to live in. For seven days, they stayed there crying, filled with sadness and regret.

After those seven days, hunger set in. They looked everywhere for food but couldn't find anything. Eve turned to Adam and said, "My dear husband, I'm so hungry. Please, go and search for something we can eat. Maybe if God sees our suffering, He'll feel sorry for us and let us return to the place we used to live."

Adam agreed and set out. For seven days he walked through the land, hoping to find food like they had in paradise, but he found nothing. When he returned, Eve said, "Why don't you end my life? Maybe if I'm gone, God will let you go back to paradise. It's my fault we were cast out."

Adam said, "Please don't say that. If you keep talking this way, God might punish us even more. I could never hurt you—you're part of me. Instead, let's go and try to find something to eat so we can stay alive."

They searched together for nine more days but still couldn't find any food like what they once had. The only food they found was what animals eat. Adam said, "This food is for animals, not us. In paradise, we ate food meant for angels. But now, the right thing to do is to ask God for mercy. Let's show we're truly sorry. Maybe He'll forgive us and help us survive."

Eve asked, "What does it mean to show we're sorry? What should I do? Let's not give ourselves something so hard that we fail. If we promise too much and can't finish it, God might turn away from us. Please tell me what you plan to do. I know I caused you pain, and I want to make it right."

Adam said, "You don't have to do the same as I do. Just do what you can. I will fast for forty days without eating. You go to the Tigris River. Stand on a stone in the water, up to your neck. Don't say anything, because we're not worthy to speak to God. Our mouths are unclean from eating what He told us not to. Stay there for thirty-seven days. I will go to the Jordan River and do the same, hoping that God will forgive us."

Eve went to the Tigris River and followed Adam's instructions. Adam went to the Jordan River, stood on a stone, and stayed in the water up to his neck.

While standing there, Adam spoke to the river. "River Jordan, feel my sadness. Gather all the creatures living in you and bring them here to mourn with me. They shouldn't cry for themselves, because they didn't sin—I did." Right away, the water stopped flowing, and all the creatures came and surrounded him.

After eighteen days, Satan became angry. He disguised himself to look like a shining angel and went to the Tigris River, where Eve was standing and crying. Pretending to care, he also cried and said, "Eve, stop crying. Don't be sad anymore. Why are you and Adam suffering like this? God heard your prayers. He has forgiven you. All the angels asked Him to have mercy, and He agreed. He sent me to bring you out of the river and give you the food you once had in paradise. Come with me—your food is waiting."

Eve believed him. She stepped out of the river, her body weak and cold from standing in the water so long. She fell to the ground, and the devil helped her up and led her back to Adam.

When Adam saw her with Satan, he cried out loudly, "Eve! What happened to your promise? How could you fall into his trap again? It was because of him that we lost our place in paradise and the happiness we had with God."

When Eve realized that the devil had tricked her into leaving the river, she fell to the ground in deep sorrow. Her crying grew louder and more painful. She shouted, "Shame on you, devil! Why are you doing this to us? What have we ever done to you? Why do you hate us so much? Did we take away your honor or make you lose your place? Why are you so filled with jealousy and evil that you chase us just to ruin our lives?"

With a heavy sigh, the devil replied, "Adam, my anger, jealousy, and sadness are all because of you. Because of you, I lost my place in heaven, where I lived in glory among the angels. I was thrown out of that high place and forced down to the earth."

Adam asked, "What are you talking about? What did I ever do to you? How have I hurt you? We didn't harm you in any way, so why are you attacking us?"

The devil answered, "Adam, don't you see? I lost everything because of you. When God made you, I was pushed out of His presence and removed from the group of angels. God gave you His breath and made you in His own image. Then Michael brought you before us and told us to bow to you. God said, 'Here is Adam. I have made him in My own image and likeness.'"

Michael went out and called all the angels. "Worship the image of God as the Lord has commanded," he said. Michael was the first to bow. Then he called me and said, "Worship Adam, the image of the Lord." I refused and said, "I will not bow to him. He's younger than I am. I existed before he was created. He should bow to me, not the other way around."

When the angels who followed me heard this, they also refused to bow. Michael warned us, "Worship Adam, the image of God, or God will be angry with you." But I said, "If God is angry with me, I'll rise above the stars and be equal to Him."

Because of this, God became angry and threw me and my followers out of heaven. We lost everything and were thrown down to the earth. We were devastated. When we saw you enjoying paradise, living in peace and beauty, we couldn't stand it. So I tricked your wife, and through her, I made sure you were also removed from your joy—just like I was cast out of mine.

When Adam heard all this, he broke down in tears and cried, "Lord, my God, my life belongs to You. Please remove this enemy who wants to destroy me. Give me the glory that he lost." Right away, the devil disappeared. But Adam stayed strong in his repentance, standing in the Jordan River for forty days.

Eve said to Adam, "May you live, my husband. Life has been given to you, because you never gave in to sin—not the first time, and not the second. But I made a mistake. I didn't obey God. Now, please send me away from your presence. I will go toward the west and live there until I die." She walked away, crying loudly in grief, and built a small shelter there. At that time, she was three months pregnant.

When it was almost time for her to give birth, Eve was in a lot of pain. She cried out to God, saying, "Please help me, Lord. Have mercy on me." But no answer came. She didn't feel God's help. She thought, "Who will tell Adam what's happening to me? I beg you, sun and stars in the sky, when you rise again in the east, please carry a message to Adam for me."

At that very moment, Adam said, "I hear Eve crying. Maybe the serpent is bothering her again." He rushed to find her and saw that she was suffering greatly. Eve looked at him and said, "From the moment I saw you, my sadness started to lift. Please pray to God for me. Ask Him to look at me with kindness and help me through this pain." So Adam prayed to the Lord for her.

Then twelve angels appeared, along with two heavenly beings, standing on either side of Eve. Michael stood on her right. He gently touched her face down to her chest and said, "Eve, you are blessed because of Adam. His prayers are strong and have been heard. That's why I've been sent to help you. Get ready—your time has come."

Eve gave birth to a son who shone brightly. Right after he was born, the baby stood up, picked a blade of grass, and gave it to his mother. They named him Cain.

Adam picked up Eve and their son and led them toward the east. Then the Lord God sent the archangel Michael to give Adam a variety of seeds. Michael also taught Adam how to plant and care for the ground so that he and his children could grow food and survive.

Later, Eve became pregnant again and gave birth to a son named Abel. Cain and Abel grew up together. One day, Eve said to Adam, "While I was sleeping, I had a dream. I saw Cain holding Abel's blood in his hands, and he was drinking it. This dream troubles me."

Adam replied, "If Cain really kills Abel, that would be terrible. Let's separate them. We'll build a home for each of them so they can live apart."

So they gave Cain the job of farming, and Abel became a shepherd. That way, they lived in different places. But even then, Cain killed Abel. At that time, Adam was 130 years old, and Abel was 122.

After Abel died, Adam and Eve had another child. They named him Seth. Adam said to Eve, "Now we have a son to take Abel's place, since Cain killed him." After Seth was born, Adam lived another 800 years. He had 30 sons and 30 daughters—63 children in total. Their descendants spread out and formed different groups across the land.

One day, Adam said to Seth, "My son, listen closely. I want to tell you something that happened after your mother and I were sent out of paradise. While we were praying, the archangel Michael, God's messenger, appeared to me. I saw a chariot like the wind, and its wheels looked like they were made of fire. Then I was lifted up into the paradise of righteousness. There, I saw the Lord sitting on a throne. His face was like blazing fire—so bright I couldn't look at it. Thousands of angels stood on both sides of His chariot.

When I saw this, I was overwhelmed with fear. I fell with my face to the ground. Then God said to me, 'You must die now because you broke My command. You listened to your wife instead of Me, even though I gave her to you so that you would guide her. But you chose to ignore My words.'

When I heard God say this, I lay flat on the ground and prayed, 'Lord, You are powerful and full of mercy. You are holy and just. Please don't let my name be erased from Your memory. Save my soul, because I'm about to die and lose my breath. Don't turn away from me—You formed me from the dust. Don't forget the one You created and cared for.'

Then God gave me a message about you. He said, 'Since you were created with a deep desire to learn, I will not take away from your children the right to serve Me. That promise will last forever.'

After hearing this, I fell down again and worshipped the Lord. I said, 'You are the everlasting and highest God. All creation praises You. You are the true Light above all other lights, the Living One, full of endless power. All spiritual beings honor You. You pour out mercy generously on all of humanity.'

As I finished worshipping, Michael took my hand and led me out of that paradise of vision, just as God had commanded. He was holding a rod, and when he touched the waters that surrounded paradise, they froze and turned solid."

After I returned, the archangel Michael came with me and brought me back to the place where he had first taken me. Listen, my son Seth, and I'll tell you the rest of what was shown to me after I ate from the tree of knowledge. I saw what would happen in this world and what God plans for all people.

God will come down surrounded by fire, and His voice will speak laws and commands. From His mouth will come a sharp sword. People will honor Him in the place where His power lives. He will show them His beautiful and holy place. There, in the land God prepares for them, they will build a temple for Him.

But even there, they will break His commands. Their temple will be burned, their land left empty, and they'll be scattered among other nations because they made God angry. Still, God will bring them back one day. They'll rebuild His house, and in the final days, it will be even greater than before.

Even so, evil will once again become stronger than good. Then God will come and live on earth in a way people can see. Goodness will shine again, and His house will be honored forever. Those who believe in Him will be protected from their enemies. God will gather a loyal people and save them forever. But those who rejected His laws will be punished by God, their King.

The skies, the earth, day and night, and all living things will obey Him and do what He says. People will no longer live in sin but will turn back to God's ways. God will reject those who do evil, but those who live right will shine like the sun in His presence.

During that time, people will be washed with water to be made clean from their sins. But those who refuse to be washed will be judged. Anyone who controls themselves and lives a good life will be blessed. On the Day of Judgment, when God shows His power, He will look closely at what everyone has done, because He is a fair and honest judge.

When Adam was 930 years old and knew his life was almost over, he said, "Call all my children. I want to bless them and speak with them before I die." They all gathered in three groups in the place where they used to worship God. They asked him, "Father, why did you call us? Why are you lying down on your bed?"

Adam said, "My children, I'm sick and in pain." They replied, "Father, what does this pain mean? What is this illness?"

Then Seth said, "Father, do you miss the fruit from paradise that you used to eat? Is that why you're so sad? If that's the case, tell me. I'll go to the nearest gate of paradise, cover myself in dust, lie on the ground, and cry out to God. Maybe He will hear me and send an angel with the fruit you're longing for."

Adam replied, "No, my son. It's not about the fruit. I feel weak, and my body hurts all over." Seth asked, "What is pain, Father? I don't understand. Please explain it to us."

Adam said, "Listen to me, my children. When God created me and your mother, He put us in paradise and gave us all the fruit trees to eat from. But He gave us one rule—we weren't allowed to eat from the tree of the knowledge of good and evil in the middle of the garden. God gave me the trees in the east and north, and your mother the trees in the south and west.

God also gave each of us an angel to watch over us. One day, the angels went up to heaven to worship. While they were gone, the devil found his chance. He tricked your mother into eating the fruit from the tree we weren't supposed to touch. She ate it and gave some to me, and I ate it too.

As soon as we disobeyed, God became angry. He said to me, 'Because you didn't follow My command, I will send pain to your body. You will suffer seventy types of pain—from your head, your eyes, your ears, down to your toes. Every part of your body will hurt. This is the punishment I've decided, not just for you, but for everyone born after you.'

After Adam said this, a strong pain hit him. He cried out, "What can I do? This pain is too much!"

When Eve saw him crying, she also began to cry. She prayed, "Lord, my God, please give me his pain instead. I'm the one who sinned." Then she said to Adam, "My husband, let me take some of your pain. This suffering came to you because of me."

Adam said to Eve, "Get up and go with our son Seth to the area near paradise. Put dust on your heads, lie down on the ground, and beg God for help. Maybe He'll have mercy and send His angel to the tree of mercy—the one that gives the oil of life. If He gives you even one drop, you can use it to touch my body so I can get some relief from this pain that's slowly destroying me." So Eve and Seth started walking toward the gates of paradise.

While they were walking, a wild animal—a serpent—suddenly came out and bit Seth. When Eve saw it happen, she cried out, "Oh no, what a miserable woman I am! I'm cursed because I didn't obey God's command." Then she yelled at the snake, "You cursed animal! How dare you attack someone made in God's image! What gives you the right?"

The serpent replied in a human voice, "Eve, we hate you most of all. You're the one we're angry at. Why did you open your mouth and eat the fruit you weren't supposed to? And now you don't even want to hear the truth."

Seth spoke firmly to the serpent: "The Lord God rebuke you! Be quiet! Stop speaking, you liar who brings nothing but trouble. Stay away from us, God's creation, until the day God brings you to justice." The serpent answered, "Fine, I'll leave like you said." And it went away, though Seth was still hurt.

Seth and Eve continued on to the edge of paradise. They put dust on their heads, lay down on the ground, and cried out to God. They begged Him to have mercy on Adam and to send His angel with the oil from the tree of mercy to ease his suffering.

After they prayed for many hours, the angel Michael appeared and said, "God sent me to you. I'm the angel who watches over people's bodies. Listen, Seth, servant of God. Don't cry or keep asking for the oil to heal Adam."

Michael continued, "You won't receive that oil until the end times. After 5,500 years, God's chosen Son, Christ, will come to earth. He will bring Adam and all the dead back to life. When He comes, He will be baptized in the Jordan River. After that, He will anoint all who believe in Him with the oil of mercy. That oil will be passed on from generation to generation for everyone who chooses new life through water and the Holy Spirit. When Christ comes, He will lead your father Adam back to paradise, to the tree of mercy."

Michael then said, "Seth, return to your father. His time is almost up. In six days, his soul will leave his body. When that happens, you'll see amazing things in the sky and on the earth." Then Michael disappeared. Seth and Eve collected sweet-smelling herbs—nard, saffron, calamus, and cinnamon—and went back.

When they reached Adam, they told him about the serpent that bit Seth. Adam said to Eve, "What have you done? You've brought sin and suffering not just to us, but to everyone who comes after us. When I'm gone, make sure you tell our children what happened. They'll struggle in life, and when things go wrong, they'll blame us and say, 'Our parents caused all of this.'" Hearing that, Eve broke down in tears.

Just as Michael had said, Adam died six days later. As he was about to take his last breath, he told his children, "I am 930 years old. When I die, bury me in the field facing the sunrise, next to this house." Then he passed away.

Right after he died, the sun, moon, and stars went dark for seven days. Seth held Adam's body and mourned deeply. Eve sat in silence with her hands over her head. All of their children cried with deep sorrow.

Then the angel Michael appeared by Adam's head and said to Seth, "Get up from your father's body and come with me. Let me show you what God has planned for him. God made him and has shown him mercy." Then all the angels blew trumpets and shouted,

"Praise to You, Lord, for showing kindness to the one You created!"

Seth saw God's hand reach down and take Adam. God gave Adam to Michael and said, "Take care of him until the Day of Judgment. In the last days, I will turn his pain into joy. He will sit on the throne of the one who took his place." Then God said to the angels Michael and Uriel, "Bring me three white linen cloths. Wrap Adam in them. Also bring cloths for Abel and bury both of them."

All the angels gathered to honor Adam. Michael and Uriel buried Adam and Abel in a part of paradise, in front of Seth and Eve. No one else was allowed to see. The angels told them, "When someone dies, bury them this way."

Six days after Adam's death, Eve knew she was going to die too. She gathered all her children—Seth, his thirty brothers, and thirty sisters—and said, "My dear children, listen carefully. I want to tell you what the angel Michael told us after your father and I disobeyed God. Because of that sin, God will judge our people. First, He'll judge by water, then later by fire. These are the two ways He'll judge all of humanity."

She continued, "Make tablets out of stone and clay. Write on them everything you've learned from your father and me—everything you've heard and seen. If God judges by water, the clay tablets will be destroyed, but the stone ones will last. If He judges by fire, the stone tablets will break, but the clay ones will become strong and survive."

After saying this, Eve raised her hands in prayer, knelt to the ground, and gave thanks to God. While she was still praying, her soul left her body. All of her children buried her with deep sorrow.

Four days later, Michael the archangel appeared and said to Seth, "Man of God, don't mourn your loved ones for more than six days.

The seventh day is a holy sign. It reminds us of the future resurrection and the rest that's coming. On the seventh day, God rested from all His work." Then Seth made the clay and stone tablets.

Ladder of Jacob

Introduction

The Ladder of Jacob is a compelling and enigmatic text, deeply embedded within the tradition of ancient Jewish mysticism and apocalyptic literature. Rooted in biblical narrative yet richly expanded by imaginative interpretation, it provides modern readers with extraordinary insight into the spiritual imagination, mystical symbolism, and theological reflection characteristic of early Jewish and Christian religious thought. Although the Ladder of Jacob is not part of the canonical scriptures, it nonetheless stands as an essential witness to the vibrant creativity, profound spirituality, and theological explorations of ancient religious communities.

The Ladder of Jacob draws its initial inspiration from a well-known passage in the Hebrew Bible (Genesis 28:10-22), in which Jacob, fleeing from his brother Esau, has a visionary experience at Bethel. In this biblical account, Jacob dreams of a ladder reaching from earth to heaven, upon which angels ascend and descend, and at the top of which God appears to him, reiterating promises previously given to Abraham and Isaac. This brief narrative, rich in symbolic imagery, captured the imagination of generations of religious interpreters, mystics, and theologians, prompting numerous expansions, interpretations, and meditations throughout Jewish and Christian traditions. Among these interpretations, the Ladder of Jacob stands out as one of the most vivid, mystical, and spiritually profound.

While its exact origins and date remain debated among scholars, the Ladder of Jacob is generally considered an early pseudepigraphal writing, likely composed in the first to third century CE. Originally written in Greek, and subsequently preserved in Slavonic manuscripts, it bridges the gap between early Jewish mysticism, apocalyptic thought, and emerging Christian spiritual traditions. The text incorporates visionary experiences, angelic dialogues, elaborate symbolic imagery, and eschatological prophecies, weaving together complex spiritual narratives that reflect contemporary concerns and spiritual aspirations of its ancient readers.

The purpose of this introduction is to equip readers with a comprehensive understanding of the historical background, literary structure, theological symbolism, and spiritual significance of the Ladder of Jacob. By exploring these dimensions thoroughly, readers will be better positioned to engage deeply with this extraordinary text, appreciating its rich symbolism, mystical insights, and enduring spiritual relevance. Despite its enigmatic nature, the Ladder of Jacob offers profound reflections on human spirituality, divine revelation, and the intricate interplay between earthly realities and heavenly mysteries.

Historical Context and Literary Origins

To fully appreciate the Ladder of Jacob, it is essential first to understand the historical and religious contexts in which it emerged. The late Second Temple period (circa 200 BCE – 70 CE) and the early centuries thereafter represent a vibrant and dynamic phase in Jewish religious and spiritual life. During this period, numerous apocalyptic and mystical writings emerged, reflecting intense theological reflection, spiritual creativity, and cultural interaction. These writings, characterized by vivid visions, angelic visitations, eschatological prophecies, and symbolic narratives, addressed profound questions concerning human destiny, divine intervention, cosmic order, and spiritual transformation.

The Ladder of Jacob belongs squarely within this literary and spiritual milieu. Although exact dating remains uncertain, most scholarly consensus places the text's composition sometime between the late first and third century CE, likely in a Jewish context strongly influenced by apocalyptic thought and emerging mystical traditions. Its initial composition in Greek indicates its origins within the broader Hellenistic Jewish world, where cultural and religious interactions significantly influenced Jewish spiritual literature.

Though preserved primarily through Slavonic manuscripts, the original Greek version of the Ladder of Jacob likely circulated widely among diverse religious communities, influencing both Jewish mystical traditions and early Christian spiritual reflections. The Slavonic preservation reflects later transmission into Eastern Christian traditions, where apocalyptic and mystical writings were highly valued and extensively copied. This textual transmission further underscores the profound cross-cultural appeal of the Ladder of Jacob, reflecting its enduring spiritual relevance and theological resonance across multiple religious traditions.

Historically, texts such as the Ladder of Jacob emerged in contexts characterized by intense religious and social upheaval, including the destruction of the Second Temple (70 CE), subsequent Jewish diaspora, and ongoing theological debates concerning messianic expectations, divine intervention, and spiritual revelation. Within this tumultuous environment, visionary literature offered solace, spiritual guidance, and theological insight, providing communities with profound reflections on divine purpose, human destiny, and cosmic realities. Thus, the Ladder of Jacob can be understood as both a theological response to historical realities and a deeply spiritual meditation on timeless questions concerning humanity's relationship with the divine.

Mystical Symbolism, Visionary Imagery, and Theological Themes

Central to the spiritual appeal and theological significance of the Ladder of Jacob are its powerful mystical symbolism and vivid visionary imagery. While drawing initial inspiration from Jacob's dream at Bethel, the text radically expands this biblical narrative, introducing detailed descriptions of angelic beings, cosmic phenomena, eschatological revelations, and profound spiritual dialogues. These expansions not only enrich the original narrative but also reveal deeper theological and mystical insights concerning divine revelation, human spirituality, and cosmic order.

One of the text's most striking symbolic features is the ladder itself—representing the profound connection between earthly and heavenly realms. The ladder functions simultaneously as a bridge, a symbol of spiritual ascent, and a profound metaphor for divine-human interaction. Its imagery reflects deep mystical conceptions of spiritual progress, human transformation, and divine revelation, emphasizing humanity's capacity to ascend spiritually toward divine realms through visionary experiences, ethical purity, and spiritual discipline. The ascending and descending angels further symbolize ongoing divine involvement in human affairs, underscoring a dynamic interplay between spiritual realities and earthly existence.

In addition to angelic visions and mystical ascents, the Ladder of Jacob introduces unique theological themes concerning the divine form, often depicting God and angelic beings through elaborate symbolic descriptions, including visions of majestic divine figures, celestial mysteries, and profound spiritual truths. Such depictions echo broader Jewish mystical traditions, anticipating themes developed further in later Jewish mystical texts such as the Merkabah and Kabbalistic literature. These mystical portrayals reveal a deep fascination with divine transcendence, cosmic majesty, and spiritual illumination, reflecting theological concerns central to Jewish and Christian mysticism alike.

Furthermore, the text incorporates significant eschatological and prophetic themes, offering visionary insights into future events, divine judgment, messianic expectations, and ultimate cosmic renewal. These prophetic visions provide readers with powerful reflections on divine justice, human destiny, and cosmic redemption, reflecting broader apocalyptic traditions prevalent within Second Temple Judaism and early Christianity. By exploring such themes, the Ladder of Jacob addresses profound existential questions concerning suffering, justice, hope, and ultimate fulfillment, offering spiritual reassurance and theological guidance to its readers.

Legacy, Influence, and Modern Spiritual Relevance

Despite its exclusion from canonical scriptures, the Ladder of Jacob significantly impacted religious and mystical thought, influencing both Jewish and Christian mystical traditions through its symbolic richness and profound theological insights. Historically, the text's vivid imagery, mystical symbolism, and eschatological themes provided essential resources for later religious literature, influencing subsequent mystical traditions, apocalyptic writings, and spiritual reflections. Its preservation within Slavonic manuscripts underscores its ongoing importance within Eastern Christian spiritual traditions, highlighting its enduring resonance across diverse religious communities.

Today, the Ladder of Jacob continues to attract scholarly and spiritual interest, offering modern readers valuable insights into ancient spiritual traditions, theological reflection, and mystical imagination. Contemporary scholarly efforts increasingly appreciate the text's significance within historical contexts, religious dialogues, and comparative mystical studies, recognizing its vital role in enriching modern understandings of Jewish and Christian spirituality.

For contemporary spiritual seekers, the text offers profound insights into human spirituality, divine revelation, and the transformative power of mystical experience. Its emphasis on spiritual ascent, ethical purity, visionary revelation, and divine-human interaction resonates deeply with modern spiritual concerns, highlighting humanity's ongoing quest for spiritual fulfillment, divine encounter, and transcendent meaning.

In conclusion, the Ladder of Jacob stands as a remarkable testament to the enduring power, richness, and complexity of ancient religious thought, offering profound spiritual reflections, theological insights, and mystical imagery. By engaging deeply with this extraordinary text, readers today are invited into a profound dialogue with ancient spiritual traditions, theological creativity, and mystical symbolism, enriching contemporary understandings of religious experience, spiritual transformation, and divine revelation. The Ladder of Jacob thus remains an essential resource for exploring humanity's timeless spiritual quest, illuminating the enduring mysteries of divine encounter, cosmic purpose, and spiritual ascent.

Ladder of Jacob

Jacob went to stay with his uncle Laban. As the sun set, he found a spot to sleep, using a stone as a pillow. That night, he had a dream.

In the dream, he saw a ladder standing on the ground that reached all the way to the sky. At the top of the ladder was a glowing face that looked like it was made of fire and shaped like a person. The ladder had twelve steps, and on each step were two human faces—one on the left and one on the right—making twenty-four faces in total. But one face in the middle was bigger and brighter than the rest. It stretched from the shoulders to the arms and looked more powerful and frightening than any of the others.

As Jacob watched, he saw angels going up and down the ladder. Then he looked up and saw the Lord standing above it. The Lord called out, "Jacob, Jacob!" Jacob replied, "I'm here, Lord." Then the Lord said, "The land you are lying on will belong to you and your children after you. I will make your family as countless as the stars in the sky and the sand by the sea. Because of your family, all the people on earth will be blessed—even in the far future. The blessing I give you will continue for many generations. People from the east and west will be part of your family."

Hearing this made Jacob shake with fear. He woke up, still feeling the presence of God's voice. He said, "This place is holy! It must be God's house and the gate of heaven." Then he took the stone he had used as a pillow, stood it upright as a marker, poured oil over it, and named the place "God's House."

Then he prayed, "Lord God, Creator of Adam, God of Abraham and Isaac, my father, and of all who live faithfully—you sit above the cherubim and rule from a throne that glows with fire and is covered in eyes. You carry the cherubim with four faces, you support the seraphim with many eyes, and you hold the world in your arms while no one holds you. You created the heavens to honor your name and spread the skies above the clouds. You guide the sun and hide it at night so people don't think it's a god. You made a path for the moon and stars. The moon grows and shrinks, and the stars move just as you told them to—so they won't be mistaken for gods either.

The six-winged seraphim stand in front of you. They cover their faces and feet with their wings and fly with the others, singing, 'Holy, Holy, Holy!' You are the Most High with many names. You are fire and lightning, full of glory. You are Jao, Jaoel, Sabakdos, Chabod, Sabaoth, Omlelech, Elaber, and more. You are the eternal King—strong, mighty, patient, and full of greatness. You fill the heavens, the earth, the sea, and even the deepest places with your glory. Please hear my prayer. Listen to my praise. Show me what my dream means. You are strong, holy, and full of glory. You are my God and the God of my fathers."

While Jacob was still praying, a voice appeared in front of him. It said, "Sarekl, leader of those who serve with joy and keeper of visions—go help Jacob understand the dream he saw. Show him everything he saw, but first, bless him."

Then the archangel Sarekl came to Jacob. His face looked powerful, but Jacob wasn't scared. The face he had seen in the dream was even more intense, so he wasn't afraid of the angel. The angel asked, "What is your name?" Jacob answered, "Jacob." Then the angel said, "From now on, you will no longer be called Jacob. Your name will be like mine—Israel."

Later, as Jacob traveled from Fandana in Syria to meet his brother Esau, Esau came to him, blessed him, and also called him Israel. Esau didn't tell Jacob his own name at first. Only when Jacob begged him did Esau finally share it and explained that it was connected to something Jacob had done. (This part seems to mix up Jacob's dream with the story of when he wrestled the angel, and the details are a little unclear.)

The angel explained, "The ladder you saw with twelve steps and faces on each side stands for the world you're living in now. Each step is a period of time, and the twenty-four faces are the kings who rule over non-believing nations. During their rule, your descendants will suffer because of their sins. The place you saw will be destroyed four times. Still, the temple will be rebuilt to honor the God of your ancestors. But because of their sin, it will be ruined again, and will stay that way until the fourth generation has passed. That's why you saw four visions."

Then Jacob saw someone trip while climbing the ladder. He saw angels moving up and down and more faces along the steps. God would raise a leader from Esau's family. All the rulers of the nations who had hurt Jacob's people would give in to this leader. He would treat them harshly and rule over them with force. They wouldn't be able to stop him. Eventually, he would demand that people worship false gods and the dead. Many from Jacob's family would be handed over—some to this harsh ruler, and others to God's judgment.

Jacob, I want you to understand this: your descendants will one day live in a foreign land, where they will be treated badly and forced into slavery. They will be beaten and hurt daily. But the people who treat them this way will be judged by God. When the time comes and a king rises up and fights back, God will bring judgment to that place. Then your people, the children of Israel, will be freed from the power of those who ruled over them with cruelty. They will no longer be shamed or mocked by their enemies.

This king will be the one who brings justice and punishment to those who attacked Israel. At the end of this time, those who suffer will cry out, and the Lord will listen. He will feel compassion, and even the strongest will show pity because angels and archangels will pray for the rescue of your people. Women among your people will have many children again, and the Lord will fight for them.

The land of their enemies will suffer. Their food supplies will be empty, there will be no wine or fruit. The ground will be full of crawling creatures and other harmful things. There will be earthquakes and destruction. When God passes judgment on that land, He will lead your people out of slavery. They will be saved from the insults of their enemies.

The king who stood against them will face revenge. Even though he stood proudly, thinking he was powerful, the people cried out—and the Lord heard them. He poured out His anger on Leviathan, the sea monster, and struck down the wicked ruler named Thalkon with a sword because that ruler had lifted himself up against the true God.

Then, Jacob, your righteousness and that of your ancestors—and of your children who follow your path—will be seen. Your descendants will blow the trumpet, and the entire kingdom of Edom will be destroyed, along with the kings and people of Moab.

The dream you had about the angels going up and down the ladder points to something in the future. In the last days, someone will come from God. He will want to connect what is above with what is below. Before He arrives, your sons and daughters will speak messages from God, and young people will have visions of Him.

Strange and powerful signs will happen when He is about to come: a tree that is cut down will bleed, babies just a few months old will speak clearly, and a child still in the womb will announce His coming. A young man will seem wise like an old man. Then, the One who has long been expected will arrive, but no one will know exactly how He comes. When He appears, the earth will celebrate because heaven's glory has come down to it. What was once far above will now be close.

From your family, a royal leader will grow. He will rise up and destroy the power of evil. He will be a savior—not just for your people, but for the tired and broken among the other nations too. He will be like a cloud offering shade from the heat, covering the whole world. Without Him, the world would remain broken. He is the one who connects heaven and earth.

When He arrives, idols made of bronze, stone, and carved images will speak out for three days. They will tell wise people what is happening on the earth. And by a special star, those seeking Him will find their way. They will see Him walking on earth—this One whom even the angels cannot fully see in heaven.

Then, God Himself will appear in human form. He will be held in the arms of a regular person. He will bring new life to humankind. He will restore what was lost when Adam and Eve sinned. The lies and tricks of evil will be broken, and all false gods will fall down in shame because they were built on lies. They will no longer be able to rule or pretend to speak truth. Their power will be taken from them, and they will have no more honor.

The child who comes will take their strength away. He will fulfill the promise God made to Abraham. He will smooth out everything rough and bring peace. He will throw all wickedness into the deep sea and perform amazing miracles in both heaven and earth.

But this savior will be wounded—in the house of someone He loves. When He is hurt, the time of salvation and the end of all evil will be near. Those who hurt Him will be wounded in return, and their wound will never heal. But the One who was wounded will be worshiped by all creation. People from everywhere, including all nations, will place their hope in Him. Everyone who knows His name will never be ashamed. His strength and His life will never end.

Joseph and Asenath

Introduction to Joseph and Asenath

Joseph and Asenath is a fascinating story that expands on the brief mention in the Bible of Joseph's marriage to Asenath, the daughter of an Egyptian priest, as told in the book of Genesis. This text, often linked to the Apocrypha, explores their relationship in detail, focusing on Asenath's journey as she accepts Joseph's faith and how their love story unfolds.

The story combines Jewish and Hellenistic writing styles and explores themes like repentance, divine guidance, and the blending of different cultures and beliefs. Likely written during the Second Temple period, it provides insight into the interactions between Judaism and the Greco-Roman world at that time.

Including this story in religious literature showcases the depth of ancient storytelling and interpretation. It also reinforces the lasting impact of Joseph's story as a model of faith, righteousness, and the power of reconciliation.

Chapter One

Pharaoh gave Joseph the name Zaphnath-Paaneah because of his ability to reveal secrets. He also gave him Asenath, the daughter of Potipherah, the priest of On, to be his wife. After this, Joseph traveled throughout Egypt.

Before the famine began, Asenath gave birth to two sons. Joseph named them Manasseh and Ephraim while living in Egypt.

In the first year of the seven years of abundance, during the second month, Pharaoh sent Joseph on a journey across the land of Egypt. By the fourth month of that year, on the eighteenth day, Joseph arrived in Heliopolis, where he started gathering all the grain from the land. The harvest was so vast that it was as countless as the sand on the seashore.

In Heliopolis, there was a man named Pentephres, who was one of Pharaoh's most important officials. He was a satrap (a high-ranking leader) and the chief priest of Heliopolis. He was extremely wealthy, wise, and generous, serving as one of Pharaoh's closest advisors.

Pentephres had a daughter named Asenath, who was about eighteen years old. She was tall, graceful, and more beautiful than any other young woman in Egypt. She did not look like the daughters of the Egyptians but resembled the Hebrew women. She was as tall as Sarah, as stunning as Rebecca, and as radiant as Rachel.

Asenath's beauty was known throughout the land, even in the farthest regions. Many young men—sons of lords, governors, and kings—wanted to marry her. Because of this, there was fierce competition, and fights broke out over who would win her hand.

When Pharaoh's oldest son heard about her beauty, he asked his father to let him marry Asenath. He said, "Give me Asenath, the daughter of Pentephres, the priest of Heliopolis, to be my wife."

But Pharaoh refused, saying, "Why would you choose a wife of lower status than yourself? Are you not the ruler of all the earth? No! You are already engaged to the daughter of King Joakim. She is a queen and very beautiful—she will be your wife instead."

Chapter Two

Asenath looked down on all men and thought they were beneath her. No man had ever seen her because her father, Pentephres, had built a tall and grand tower inside his home, where she lived. The top floor of the tower had ten rooms.

The first room was large and beautiful, with a floor made of purple stones and walls decorated with precious gems. The ceiling was covered in gold, and the room was filled with countless Egyptian idols made of gold and silver. Asenath worshiped these idols, feared them, and offered sacrifices to them.

The second room was where she kept her finest clothing and treasures. It contained chests filled with gold, silver, and luxurious garments woven with gold, as well as expensive jewels and fine linens. This room also held all her personal ornaments.

The third room was a storage area, filled with the best crops and produce from the land.

The remaining seven rooms belonged to seven young women, each having her own space. These young women were Asenath's attendants, and they were all the same age as her, having been born on the same night. They were as beautiful as the stars in the sky, and no man or boy had ever been involved with them.

Asenath's main room, where she spent most of her time, had three windows: one facing the courtyard to the east, one looking north onto the street, and another facing south. A golden bed stood in the room, positioned toward the east. It was covered with a luxurious purple blanket woven with gold, decorated with blue embroidery and fine linen. Asenath slept alone in this bed, and no one—man or woman— had ever sat on it.

Outside the tower, there was a large courtyard, surrounded by a tall wall made of massive stone blocks. The courtyard had four iron-covered gates, each guarded by eighteen strong young men.

Inside the courtyard, beautiful trees of every kind had been planted, all filled with ripe fruit, since it was harvest season.

On the right side of the courtyard, there was a spring of fresh water that never stopped flowing. Beneath it, a large cistern collected the water, which then flowed through the middle of the courtyard like a river, watering all the trees.

Chapter Three

On the eighteenth day of the fourth month, Joseph arrived in the district of Heliopolis. As he neared the city, he sent twelve men ahead to deliver a message to Pentephres, the priest of Heliopolis.

Joseph's message said, "May I stay at your house today? It is almost noon, and it is time for the midday meal. The sun is very hot, and I would appreciate some rest under your roof."

When Pentephres heard the message, he was overjoyed and said, "Blessed be the Lord, the God of Joseph." Then he called his steward and said, "Hurry! Prepare the house and make everything ready. Prepare a great feast, for Joseph, the mighty man of God, is coming to visit us today."

Meanwhile, Asenath heard that her father and mother had returned from their country estate. She was excited and said, "I will go see my father and mother, for they have come back from their estate."

She quickly dressed in her finest clothes. She put on a blue linen robe woven with gold and tied a golden belt around her waist. She adorned herself with golden bracelets on her wrists and ankles and wore golden trousers along with a gold necklace around her neck. Her jewelry was decorated with precious stones, each engraved with the names of the Egyptian gods. These names were carved into the stones of her bracelets and other ornaments.

She placed a tiara on her head, wrapped a diadem around her temples, and covered her head with a veil.

Chapter Four

Asenath quickly ran down the stairs from her room at the top of the tower and went to greet her father and mother. Pentephres and his wife were overjoyed to see their daughter, dressed beautifully like a bride of God. They gave her many gifts from their country estate, bringing out the best things they had.

Asenath was delighted with the gifts, which included fruit, grapes, dates, doves, pomegranates, and figs. Everything was fresh and delicious.

Pentephres looked at her and said, "My child."

She replied, "Here I am, my lord."

He said, "Sit here between us; I have something important to tell you." So Asenath sat down between her father and mother.

Pentephres took her right hand in his and said, "My child."

Asenath responded, "What is it, Father?"

Pentephres continued, **"Listen, Joseph, the mighty man of God, is coming to visit us today. Pharaoh has made him the ruler over all of Egypt. He is in charge of distributing grain and is saving the country from famine.

Joseph is a man who worships God. He is wise, virtuous, and, like you, he has never been with a woman. He has great knowledge, and God's spirit is with him. The Lord's grace is upon him. So, my child, I will give you to him as his wife. You will be his bride, and he will be your husband forever."**

When Asenath heard this, her face turned red with anger, and she became furious. She glared at her father and said,

**"Why are you saying this, my lord and father? Why would you treat me like a captive and give me to a man from another land? He was once a runaway and was sold as a slave!

Isn't he just a shepherd's son from Canaan, abandoned by his own family? Didn't he get accused of being with his master's wife and then thrown into prison, left in darkness until Pharaoh freed him because he explained a dream? No! I will marry the king's eldest son because he is the true ruler of the world."**

Hearing this, Pentephres decided not to say anything more about Joseph, realizing that his daughter was too proud and angry to listen.

Chapter Five

One of Pentephres's young servants suddenly ran in and shouted, "Joseph is at the gates of our courtyard!"

As soon as Asenath heard this, she quickly left her father and mother, ran upstairs to her room, and stood at the large window facing east. From there, she watched as Joseph entered her father's house.

Pentephres, his wife, and all their relatives went out to welcome Joseph. The eastern gates of the courtyard were opened, and Joseph entered, riding in Pharaoh's chariot of honor. The chariot was pulled by four white horses, as pure as snow, with golden reins, and it was covered in shining gold.

Joseph was dressed in a magnificent white tunic, and over it, he wore a purple robe woven with gold. On his head was a golden crown decorated with twelve precious stones, and above them were twelve golden rays shining like the sun. In his right hand, he held a royal scepter.

Joseph also carried an olive branch, full of ripe fruit.

As Joseph entered the courtyard, the gates were shut behind him. No strangers—neither men nor women—were allowed inside, and the gatekeepers locked the doors.

Pentephres, his wife, and all their relatives—except Asenath—stepped forward and bowed down to Joseph, pressing their faces to the ground.

Joseph then stepped down from his chariot and reached out his right hand to greet them.

Chapter Six

Asenath saw Joseph and was completely overwhelmed. Her stomach turned, her knees felt weak, and her whole body trembled. Fear filled her heart, and she cried out,

**"Where can I go? Where can I hide from him? How will Joseph, the son of God, even look at me, knowing the terrible things I said about him?

Where can I run and hide when he sees everything, and no secret is hidden from him because of the great light within him? May Joseph's God have mercy on me, for I spoke against him without understanding the truth.

What hope is there for me now? How foolish I was! Didn't I say that Joseph was just a shepherd's son from Canaan? But now I see the truth—he shines like the sun, and today, he has entered our home in all his glory.

I was reckless and arrogant to speak badly about him. I didn't realize that Joseph is truly the son of God.

What man could ever be born with such beauty, and what mother could give birth to someone so radiant? I was blind and foolish to insult him in front of my father.

Now, let my father give me to Joseph as a servant and a slave, and I will serve him for the rest of my life."**

Chapter Seven

Joseph entered Pentephres's house and sat down. After washing his feet, a separate table was set before him because he would not eat with the Egyptians, as their customs were different from his own.

Looking toward the window, Joseph asked, "Who is that woman standing in the sunlight? Tell her to leave."

He said this because he was cautious. Many of the wives and daughters of Egypt's nobles and officials had tried to win his attention. Because Joseph was very handsome, they often longed for him, and some even sent messengers with gold, silver, and expensive gifts, hoping he would notice them.

But Joseph always refused, saying, "I will not sin against the God of Israel." He kept his father Jacob's teachings in his heart, remembering the words his father often told him and his brothers:

"My sons, stay away from immoral women, for they lead to destruction."

That was why Joseph said, "Tell that woman to leave."

Pentephres answered, "My lord, the woman you saw is not a stranger. She is our daughter, a virgin who wants nothing to do with men. No man has ever seen her—until today. If you wish, she can come speak with you, for she is like your sister."

When Joseph heard this, he was pleased and said, "If she is your daughter, let her come. From today, I will consider her my sister."

Chapter Eight

Asenath's mother went upstairs to her room and brought her down to meet Joseph. Pentephres said to his daughter, "Greet your brother, for he is also pure, just as you are today. He avoids foreign women, just as you avoid foreign men."

Asenath turned to Joseph and said, "May you be joyful, my lord, for you are blessed by God Most High."

Joseph replied, "May God, who gives life to everything, bless you."

Then Pentephres said to Asenath, "Come closer and kiss your brother."

As Asenath stepped forward to kiss Joseph, he held out his right hand, placed it on her chest, and stopped her. He said,

**"It is not right for a man who worships God, who praises the living God with his mouth, eats the blessed bread of life, drinks the holy cup of immortality, and is anointed with sacred oil, to kiss a woman who blesses lifeless idols with her mouth.

A woman like that eats from their table—the bread of sorrow—drinks from their cups, which lead to betrayal, and is anointed with oil that brings destruction.

A man of God may kiss his mother, his sister from his own people, or the wife who shares his life, for these women bless the living God with their mouths. In the same way, a woman who follows God should never kiss a foreign man, for this is unacceptable in God's eyes."**

When Asenath heard Joseph's words, she was heartbroken. She began to cry loudly, and tears filled her eyes as she looked at him.

Joseph saw her sadness and felt compassion for her. He was kind, merciful, and full of reverence for God.

He lifted his right hand over her head and prayed,

**"O Lord, the God of my father Israel, the Most High, the Mighty One, who gives life to all things, who calls them out of darkness into light, out of confusion into truth, and out of death into life, bless and renew this young woman with Your Spirit.

Transform her with Your unseen hand and give her new life. May she eat the bread of life and drink from Your cup of blessing. You chose her even before she was born. May she enter the rest You have prepared for those You have chosen."**

Chapter Nine

Asenath felt overwhelming joy after Joseph blessed her. She quickly went up to her room at the top of the tower and collapsed onto her couch, completely exhausted. While she was happy, she also felt uneasy and deeply afraid. Her whole body was covered in sweat from the moment Joseph spoke to her in the name of God Most High.

She began to cry bitterly and regretted worshiping her old gods. She waited for evening to come, lost in her thoughts.

Meanwhile, Joseph finished eating and drinking, then said to his servants, "Hitch the horses to the chariot. I must leave now and travel through the city and the surrounding area."

Pentephres turned to Joseph and said, "Please, my lord, stay the night with us and leave in the morning."

But Joseph replied, "No, I must go now, for today is the day when God first began His works. In eight days, I will return and stay the night with you."

Chapter Ten

Then Pentephres and his family returned to their estate, leaving Asenath alone with her attendants. But Asenath felt restless and cried until sunset. She refused to eat or drink, and while everyone else slept, she stayed awake.

She quietly opened the door and went down to the gate, where she found the gatekeeper asleep with her children. Moving quickly, she took down the leather curtain from the door, filled it with ashes, carried it back to her room at the top of the tower, and spread the ashes across the floor. She then locked the door with an iron bar and began weeping loudly.

One of the young women who was closest to Asenath heard her sobbing. She woke the others, and together they went to Asenath's door, but they found it locked. Hearing her cries, they called out, "Why are you so sad, my lady? What's wrong? Open the door so we can help you."

Asenath answered from inside, "I have a terrible headache and am lying down. I don't have the strength to open the door because I'm so exhausted. Please go back to your rooms."

Later, Asenath got up, quietly unlocked her door, and went into her second room, where she kept her fine clothes and treasures. She opened her wardrobe and took out a black mourning tunic, the one she had worn when her oldest brother died. She removed her royal robe and put on the black tunic. She untied her golden belt and replaced it with a simple rope around her waist. She also took off her tiara, diadem, and bracelets.

Then, she picked up her finest royal robe and threw it out the window for the poor. Next, she gathered all her gold and silver idols, broke them into small pieces, and tossed them outside for the needy.

Asenath also took her royal feast—the meats, fish, and offerings meant for her gods—along with the wine vessels used in their ceremonies, and threw them out the window as food for the dogs. Afterward, she spread the ashes all over her floor.

She put on sackcloth, tied it around her waist, took the band from her hair, and sprinkled ashes over herself. Then she lay down in the ashes, beating her chest and crying bitterly through the night.

By morning, Asenath saw that the ashes beneath her were soaked with tears, turning them into mud. She fell face down into them again and stayed there until sunset.

Asenath continued this for seven days, refusing to eat or drink.

Chapter Eleven

On the eighth day, Asenath lifted her head from the floor where she had been lying. Her body had become weak from the many days of suffering.

Chapter Twelve

Aseneth reached out her hands toward the east and looked up to the sky, saying:

"O Lord, God of all time, You gave life to everything. You brought things that were hidden into the light and created all that exists. You raised the heavens and placed the earth upon the waters. You set the great stones in the sea so they would not sink but remain as You commanded.

O Lord, my God, I call out to You. Please hear me, and I will admit all my sins. I will confess everything I have done wrong.

I have sinned, Lord—I have disobeyed Your law and done evil. I have spoken wicked things in Your presence. My mouth has been made unclean by eating food offered to idols and by sitting at the table of the Egyptian gods.

I have worshiped lifeless idols that cannot hear or speak. I do not deserve to speak to You, for I am full of shame. I, Aseneth, daughter of Pentephres the priest, was proud and arrogant, and I have done wrong.

Now, Lord, I lift my prayer to You and cry out for help. Save me from those who want to harm me! I run to You like a child seeking protection from their parents.

Reach out to me, Lord, like a father who loves and protects his children. Rescue me from the hands of my enemy.

A fierce and ancient lion is chasing me—his children are the Egyptian gods I rejected and destroyed. Their father, the Devil, wants to destroy me.

But You, Lord, are my only hope. Save me from his grip! Do not let him take me like a wolf catching its prey. Do not let him throw me into fire or deep waters. Do not let the great sea creature swallow me whole.

Lord, I am alone. My father and mother have turned away from me because I destroyed their gods. But I trust in You alone. You are the Father of orphans, the protector of those who are hunted, and the helper of those who suffer.

All my father's gods are weak and temporary, but those who belong to You, Lord, will last forever."

Chapter Thirteen

Look at me, Lord—I am alone, with no family to turn to, so I have come to You for shelter.

I have taken off my royal robe, woven with gold, and put on a simple black garment instead. I removed my golden belt and tied a rough rope and sackcloth around my waist. I took the crown from my head and covered myself in ashes.

The floor of my room, once decorated with colorful stones and purple fabrics and filled with the scent of myrrh, is now soaked with my tears and covered in ashes. The mixture of my tears and ashes has turned into mud, making my room look like a street walked by many.

I gave away my royal feast and the finest meats to the dogs. For seven days and nights, I have eaten nothing and drunk no water. My mouth is as dry as a drum, my tongue stiff like a horn, my lips cracked like broken pottery. My face is thin, and my eyes are weak from crying nonstop.

But, Lord, please forgive me. I sinned without knowing the truth. I spoke wrongly about my lord Joseph because I did not realize who he really was.

I was told he was just the son of a shepherd from Canaan, and I believed it. I was blind to the truth and looked down on him, not knowing that he is Your chosen one.

Who has ever been as handsome as Joseph? Who else is as wise and strong? But now, Lord, I place him in Your hands, for I love him more than my own life.

Please keep him in Your wisdom and grace. And if it is Your will, let me serve him—I will wash his feet, care for him, and be his servant for the rest of my days.

Chapter Fourteen

As Aseneth finished her prayer and confessed her sins to the Lord, the morning star rose in the eastern sky.

When she saw it, joy filled her heart, and she said, "The Lord has heard me! This star is His messenger, bringing news of a new and glorious day."

Suddenly, the sky around the morning star opened up, and a brilliant light appeared, unlike anything she had ever seen.

Overcome with awe, Aseneth fell face-down into the ashes. Then, a man came down from heaven and stood beside her.

He called out, "Aseneth."

She was startled and said, "Who is calling me? My door is shut, and my tower is high—how could anyone have entered my room?"

The man spoke again, "Aseneth, Aseneth."

She replied, "Here I am, my lord. Please, tell me who you are."

He answered, "I am the leader of the Lord's house, the commander of all His armies. Stand up, and I will speak to you."

Aseneth lifted her head and saw a man who looked exactly like Joseph. He wore a royal robe, a crown, and carried a staff.

But his face shone like lightning, his eyes blazed like the sun, his hair looked like burning fire, and his hands and feet glowed like heated iron.

Aseneth trembled with fear and fell to the ground at his feet.

The man said, "Do not be afraid, Aseneth. Stand up and listen to my words."

She rose to her feet, and he continued, "Take off the black robe you are wearing and remove the sackcloth from your waist. Brush the ashes from your head and wash your face with water.

Put on a new robe, one that has never been worn before, and fasten the bright double belt of your purity around your waist."

Then he added, "When you have done this, return to me, and I will share the message I was sent to give you."

Aseneth went into the room where she kept her fine clothes and treasures. She opened her wardrobe, took out a beautiful, clean robe, and replaced her black tunic with it.

She untied the rough rope and sackcloth from her waist and fastened the shining double belt of her purity—one piece around her waist and the other across her chest.

Then, she shook the ashes from her head, washed her face with pure water, and covered her head with a fine, elegant veil.

Chapter Fifteen

Aseneth returned to the man, and when he saw her, he said, "Take off your veil, for today you are pure, and your head is like that of a young person."

She removed her veil, and the man continued, "Be strong, Aseneth, for the Lord has heard your prayers and confession.

Do not be afraid, for your name is now written in the book of life, and it will never be removed. From this day on, you are made new, given a fresh start, and blessed with a new life. You will eat the bread of life, drink from the cup of eternity, and be anointed with oil that never fades.

Take heart, Aseneth, for the Lord has chosen you to be Joseph's wife, and he will be your husband.

Your name will no longer be Aseneth but 'City of Refuge,' because many nations will find safety in you. Under your protection, many people will take shelter, and within your walls, those who turn to God will find peace.

For Penitence, the daughter of the Most High, is always pleading for you and for everyone who repents. The Most High is her Father, and she is the mother of those who remain pure. She prays every moment for those who turn back to God and has prepared a heavenly wedding chamber for those who love her. She watches over them forever.

Penitence is pure, beautiful, and gentle. The Most High treasures her, and all His angels honor her.

Now, I am going to Joseph to speak to him about you. He will come to see you today, and he will rejoice because you will be his bride.

Listen carefully, Aseneth: put on your wedding dress—the special robe that was kept in your room. Wear your finest jewelry and prepare yourself as a bride, for Joseph is coming to see you today, and when he does, he will be filled with joy."

When the man finished speaking, Aseneth's heart overflowed with happiness.

She fell at his feet and said, "Blessed is the Lord who sent you to rescue me from darkness and bring me into the light. May His name be praised forever.

Please, my lord, if I have found favor with you, stay a little while. Sit on my bed, and I will prepare a meal for you. I will bring you the finest wine before you continue on your journey."

Chapter Sixteen

The man said to her, "Please bring me a honeycomb as well."

Aseneth replied, "Let me send someone to my family's estate in the countryside to bring one for you, my lord."

The man said, "Go into your inner room, and you will find a honeycomb there."

Aseneth went into her inner room and, just as he said, found a honeycomb sitting on the table. It was pure white, overflowing with honey, and its scent was sweet and fresh, like the breath of life itself.

She picked it up and brought it to the man. He looked at her and said, "Why did you say there was no honeycomb in your house? Yet now you have brought me this."

Aseneth answered, "My lord, I truly had no honeycomb here before. It appeared just as you said. Could it have come from you? Its fragrance is as rich as myrrh."

The man placed his hand on her head and said, "You are blessed, Aseneth, because God's mysteries have been revealed to you. Blessed also are those who turn to the Lord in repentance, for they will eat from this honeycomb.

The bees of Paradise have made this honey, and the angels of God eat from it. Anyone who eats it will never die."

Then the man stretched out his right hand, broke off a piece of the honeycomb, and ate it. He took another piece and placed it in Aseneth's mouth.

Next, he reached out his hand and touched the edge of the honeycomb that faced east. The path of his finger turned red, like blood.

Then he stretched his hand again and touched the edge facing north. Once more, the trail of his finger turned to blood.

Aseneth stood beside him, watching everything.

Suddenly, bees began to emerge from the honeycomb's cells. They were as white as snow, and their wings shimmered with shades of purple, blue, and gold. Tiny golden crowns rested on their heads, and they had sharp stingers.

The bees flew in circles around Aseneth, from her feet to her head. Larger bees, nearly the size of queen bees, gently rested on her lips.

The man spoke to them, "Go now to your places."

Immediately, the bees left Aseneth, fell to the ground, and lay still.

Then the man said, "Get up now and return to your places."

At his words, the bees rose up and flew away, each one heading toward the courtyard around Aseneth's tower.

Chapter Seventeen

The man looked at Aseneth and said, "Did you see that?"

She replied, "Yes, my lord, I saw everything."

He nodded and said, "Just as you have seen, so it will be with the words I have spoken to you."

Then he reached out and touched the honeycomb. Suddenly, fire rose up from the table and burned it completely. As it burned, a fresh and sweet fragrance filled the entire room.

Aseneth turned to the man and said, "My lord, I have seven young women who were raised with me and serve me. They were born on the same night as I was, and I love them like sisters. Please, let me call them so that you may bless them as you have blessed me."

The man said, "Call them."

Aseneth quickly summoned the seven virgins, and when they arrived, the man blessed them, saying, "May God, the Most High, bless you forever."

Then he turned to Aseneth and said, "Take this table away."

As she turned to move it, the man suddenly disappeared from her sight. She looked up and saw something like a fiery chariot rising into the sky, heading east.

Overcome with awe, Aseneth whispered, "Have mercy on me, Lord, for I spoke wrongly before You without understanding."

Chapter Eighteen

As this was happening, a young man, one of Joseph's servants, arrived and said, "Look! Joseph, the great man of God, is coming to see you today."

Aseneth quickly called her steward and said, "Prepare a special meal for me, because Joseph, the mighty man of God, will be visiting us."

Then she went into her room and opened her wardrobe. She took out her finest robe, which shimmered like lightning, and put it on.

Around her waist, she tied a dazzling royal belt decorated with precious stones.

She put golden bracelets on her wrists, slipped golden sandals onto her feet, and placed an expensive necklace around her neck.

On her head, she set a golden crown, its front decorated with the most precious gems.

Finally, she covered her head with a veil.

Turning to her maidservant, she said, "Bring me fresh water from the spring."

Aseneth bent over the basin made of polished shell, and as she looked into the water, her face glowed like the sun, and her eyes sparkled like the morning star at dawn.

Chapter Nineteen

A young servant came to Aseneth and said, "Joseph has arrived and is waiting at the gates of the courtyard."

Aseneth quickly went down with her seven attendants to meet him.

When Joseph saw her, he said, "Come to me, pure one, for I have received a message from heaven telling me everything about you."

Joseph reached out his hands and embraced Aseneth, and she held him tightly in return. They greeted each other with warmth and joy, their hearts filled with new life.

Chapter Twenty

Aseneth said to Joseph, "Come inside, my lord, and be welcome in my home." She gently took his right hand and led him inside.

Joseph sat down in the seat of her father, Pentephres, while Aseneth brought water to wash his feet.

Joseph said, "Let one of your attendants wash my feet."

But Aseneth shook her head and replied, "No, my lord. Your feet are as my own, and your hands are mine. No one else will do this for you."

She insisted and washed his feet herself.

Joseph then took her right hand and kissed it, and Aseneth, in return, kissed his head.

Later that day, Aseneth's parents returned from their country estate. When they saw her sitting beside Joseph, dressed in a wedding robe, they were overjoyed. They praised God and joined them for a meal, eating and drinking together.

Pentephres turned to Joseph and said, "Tomorrow, I will invite the nobles and rulers of Egypt to celebrate your wedding, and you shall take Aseneth as your wife."

Joseph replied, "First, I must speak with Pharaoh about Aseneth, for he is like a father to me. He will officially give his blessing for me to marry her."

Joseph stayed that night in Pentephres's house, but he did not sleep with Aseneth. He said, "It would not be right for a man who serves God to be with his wife before they are married."

Chapter Twenty-one

Joseph woke up early in the morning and sent a message to Pharaoh, telling him about Aseneth.

Pharaoh immediately called for Pentephres and Aseneth to come to his palace.

When he saw Aseneth, he was amazed by her beauty and said, "The Lord, the God of Joseph, will bless you. He has chosen you to be Joseph's wife. Joseph is God's firstborn son, and you will be called the daughter of the Most High. Joseph will be your husband forever."

Pharaoh then took golden crowns and placed them on their heads, saying, "May God Most High bless you and make your family strong and prosperous for all time."

He turned them toward each other, and they kissed.

Pharaoh then held a grand wedding celebration, hosting a huge feast that lasted for seven days. He invited all the important officials and rulers of Egypt to join in the festivities.

He also issued a royal decree, saying, "Anyone who works during the seven days of Joseph and Aseneth's wedding will be put to death."

After the wedding and celebrations ended, Joseph and Aseneth were together, and she became pregnant. She later gave birth to two sons, Manasseh and his younger brother, Ephraim, in Joseph's household.

Chapter Twenty-two

After the years of plenty came to an end, the seven years of famine began to spread across the land. The hunger reached many nations, and people struggled to find food.

When Jacob heard that his son Joseph was alive and ruling in Egypt, he gathered his family and all their belongings and traveled there. They arrived in Egypt during the second month, on the twenty-first day, and settled in the fertile land of Goshen, which had been set aside for them.

One day, Aseneth said to Joseph, "I want to meet your father, because your father Israel is now my father too."

Joseph smiled and said, "Then let's go together to see him."

They set out for Goshen, and as they neared their destination, Joseph's brothers came out to greet them. They bowed deeply to Joseph and Aseneth as a sign of respect.

When they arrived at Jacob's tent, he welcomed them with open arms, blessing them and kissing them with great affection.

Aseneth embraced Jacob, wrapping her arms around his neck, and kissed him as if he were her own father. Touched by her love and respect, Jacob blessed her warmly and welcomed her into his family with joy.

Later, they all sat down together for a meal, eating and drinking in celebration of their reunion. The gathering was filled with gratitude as they gave thanks to God for bringing them back together after so many years apart.

When it was time for Joseph and Aseneth to return home, Simeon and Levi traveled with them to keep them safe. Levi walked on Aseneth's right side, and Simeon on her left, making sure she was well protected.

Aseneth held Levi's hand as they walked, feeling a special connection with him. She admired him deeply because he was not only a devoted servant of God but also a prophet with great wisdom. Levi had a rare gift—he could see letters written in the heavens, revealing divine messages. He often shared these insights privately with Aseneth, offering her guidance and spiritual understanding.

Levi also had a vision of Aseneth's future, seeing that she would one day rest in the highest heaven, a place of eternal peace and glory. His gift allowed him to understand her destiny and the great blessings that awaited her as someone chosen by God. Their bond was built on faith and a shared devotion to the Lord, making their connection not only one of family but also of deep spiritual friendship.

Chapter Twenty three

As Joseph and Aseneth were passing by, Pharaoh's eldest son spotted them from the top of a wall. The moment he saw Aseneth, he was struck by her beauty and became consumed with desire for her.

Unable to control his emotions, he sent messengers to summon Simeon and Levi. When they arrived, they stood before him, waiting to hear what he had to say.

Pharaoh's son said, "I have heard of your incredible strength and skill in battle. People say no one on earth can match you. I know that with your own hands, you destroyed the city of Shechem and, using just two swords, defeated thirty thousand warriors.

I need your help. Join me without hesitation, and I will reward you with gold, silver, servants, houses, and vast lands. Stand by my side and be loyal to me, for I have been wronged by your brother Joseph. He took Aseneth as his wife, though she was meant to be mine.

Now, come with me, and together we will fight against Joseph. I will strike him down with my sword and take Aseneth as my wife. In return, I will make you my brothers and closest allies forever. But if you refuse, I will kill you both with my sword." As he spoke, he pulled out his sword and waved it in front of them.

Simeon, known for his bravery and quick temper, instantly drew his own sword from its sheath and rushed forward, ready to strike Pharaoh's son.

But Levi, who was a prophet and could see what was about to happen, acted fast. He stepped firmly on Simeon's right foot, stopping him from attacking.

Levi turned to him and said, "Why are you letting your anger take over? We are the sons of a man who follows God, and it is not right for someone who serves Him to repay evil with evil."

Then, Levi looked at Pharaoh's son and spoke calmly and with respect. "My lord, why do you speak to us like this? We are men who worship the Lord, and our father is a servant of God Most High. Our brother Joseph is loved by God. How could we ever commit such a terrible act against him?

Listen carefully—never say such things about our brother again. But if you are determined to go through with this, then know that we are ready."

With that, Simeon and Levi both drew their swords, holding them up for Pharaoh's son to see. Levi continued, "Do you see these swords? These are the very weapons the Lord used to bring justice when the men of Shechem dishonored our sister Dinah, the daughter of Israel."

When Pharaoh's son saw their swords, he was overcome with fear. His whole body shook, and he collapsed to the ground at their feet, terrified.

Levi reached out his hand, helped him up, and said, "Do not be afraid. Just make sure you never speak against our brother again."

Simeon and Levi then left him behind, still trembling with fear.

Chapter Twenty-four

Pharaoh's son was overwhelmed with desire for Aseneth, and his frustration became unbearable. His servants noticed his distress and whispered to him, "The sons of Bilhah and Zilpah, the maidservants of Jacob's wives, Leah and Rachel, dislike Joseph and Aseneth. They are jealous of them and would be more than willing to help you with whatever you plan."

Hearing this, Pharaoh's son immediately sent messengers to summon them in secret. That night, they arrived, and he greeted them, saying, "I have heard that you are strong and skilled warriors."

Gad and Dan, the eldest among them, answered, "Tell us what you need, my lord, and we will do it."

Pleased, Pharaoh's son dismissed his other servants, saying, "Leave us. I have something private to discuss with these men." Once they were alone, he began to deceive them with lies.

"I am offering you a choice—riches and power or death. Choose wisely and live. I know you are brave men who would rather die fighting than live in shame. Now is your chance to take revenge on your enemies.

I overheard Joseph speaking to my father, Pharaoh. He said, 'Dan and Gad are only the sons of maidservants and not my real brothers. I am waiting for my father to pass away so I can deal with them and their descendants. They will not share in our inheritance because they are of lower status. They were the ones who sold me to the Ishmaelites, and I will make sure they pay for it.'

Pharaoh praised Joseph, saying, 'You are right, my son. Take some of my soldiers and punish them as they once did to you. You have my full support.'"

When Gad and Dan heard this, they became furious and said, "We put our trust in you, my lord. Tell us what to do, and we will carry it out."

Pharaoh's son replied, "Tonight, I will kill my father, for he treats Joseph as if he were his own son. You must kill Joseph, and I will take Aseneth as my wife."

Gad and Dan agreed, saying, "We will do whatever you ask. We heard Joseph telling Aseneth, 'Tomorrow, go to our estate for the grape harvest.' He has arranged for six hundred armed soldiers and fifty outrunners to escort her."

When Pharaoh's son heard this, he formed a plan. He placed each of the four men in charge of five hundred soldiers, making them officers and commanders.

Gad and Dan then said, "We will go at night and hide by the brook in the woods. You, my lord, should take fifty horsemen with bows and ride ahead. When Aseneth arrives, she will fall into our trap, and we will ambush her guards.

She will try to escape in her chariot, and that is when you will capture her. Once she is in your hands, we will kill Joseph while he is grieving over her. Then we will kill his children right in front of him."

Pharaoh's son was thrilled with this plan and immediately sent two thousand soldiers to join them. The men reached the brook and hid in the forest along the banks. Five hundred soldiers took position at the front, with a road running between them, setting the perfect trap.

Chapter Twenty-five

Pharaoh's son went to his father's chamber, planning to kill him, but the guards at the entrance refused to let him in.

"I want to see my father before I leave to check on my new vineyards," he told them.

The guards responded, "Your father is not well. He has been in pain all night and is finally resting. He gave us strict orders: 'Do not let anyone enter, not even my eldest son.'"

Frustrated and angry, Pharaoh's son stormed off. Gathering fifty mounted archers, he led them ahead, just as Dan and Gad had suggested.

Meanwhile, Naphtali and Asher confronted their brothers Dan and Gad. "Why are you plotting against our father Israel and our brother Joseph?" they asked. "Don't you see that God protects Joseph like the most precious thing in His sight?

You already sold him as a slave once before, and look at him now—he rules over the land, saves nations, and provides us with food during this famine.

If you try to harm him again, he will surely call upon the God of Israel. The Lord will send fire from heaven to destroy you, and His angels will rise up to fight against you."

Dan and Gad, furious at their brothers' words, snapped back, "Do you expect us to sit here and die like cowards? Never!"

Ignoring Naphtali and Asher's warnings, they pressed forward with their plan, moving ahead to attack Joseph and Aseneth.

Chapter Twenty-six

Aseneth woke up early in the morning and said to Joseph, "I'm going to our estate in the countryside, but I feel afraid because you won't be there with me."

Joseph reassured her, "Be strong and don't be afraid. Go with confidence, for the Lord is with you. He will watch over you and keep you safe, just as He protects what is most precious to Him.

I must stay here to distribute grain and make sure everyone in the city gets food so that no one dies from the famine in Egypt."

With that, Aseneth set off on her journey, while Joseph went to oversee the grain supply.

As Aseneth and her six hundred soldiers reached the brook, Pharaoh's son and his men suddenly launched their ambush. They jumped out from their hiding places and attacked Aseneth's guards.

A violent battle broke out, and many of her soldiers were killed. The attackers struck down all her outrunners with their swords.

Seeing the bloodshed, Aseneth panicked and fled in her chariot, racing as fast as she could to escape the chaos.

Meanwhile, Levi, the son of Leah, sensed the danger—being a prophet, he had foreseen the attack. He quickly warned his brothers, and they immediately prepared for battle. Each one grabbed a sword, strapped on a shield, and took up a spear before rushing off at full speed to rescue her.

As Aseneth fled, Pharaoh's son and fifty of his men suddenly appeared in front of her. The moment she saw them, her heart pounded with fear, and she began to tremble, unsure of what would happen next.

Chapter Twenty-seven

Benjamin sat beside Aseneth in the chariot.

He was a strong and handsome young man, about eighteen years old, with the strength of a young lion. His beauty was beyond words, but most importantly, he was a man who feared God.

As the danger grew near, Benjamin sprang from the chariot. He spotted a round stone by the brook, picked it up, and threw it with all his might at Pharaoh's son. The stone struck him hard on the left temple, leaving him badly injured. He fell from his horse, barely clinging to life.

Benjamin then climbed onto a nearby rock and called out to Aseneth's chariot driver, "Bring me fifty stones from the brook." The driver quickly gathered the stones and handed them over.

One by one, Benjamin hurled the stones at Pharaoh's son's fifty men. Every throw hit its mark, striking each soldier in the temple and killing them instantly.

At the same time, the sons of Leah—Reuben, Simeon, Levi, Judah, Issachar, and Zebulun—rushed into the ambush. They attacked with such speed and power that they defeated all two thousand enemy soldiers, wiping them out completely.

Meanwhile, the sons of Bilhah and Zilpah, who had helped with the plot, saw what was happening and ran in fear. As they fled, they shouted to each other, "We're doomed because of our brothers! Pharaoh's son has been killed by Benjamin, and all his soldiers have died at his hands. We must act fast—let's kill Aseneth and Benjamin before escaping into the woods."

With their swords still dripping with blood, they charged toward Aseneth. When she saw them coming, she cried out in desperation, "O Lord, my God, You saved me from death and promised that my soul would live forever. Please rescue me from these men!"

The Lord heard Aseneth's prayer. In an instant, the swords of her attackers fell from their hands, crumbling into dust as they hit the ground.

Chapter Twenty-eight

The sons of Bilhah and Zilpah saw what had happened and were filled with fear. They turned to each other and said, "The Lord Himself is fighting for Aseneth against us."

Terrified, they fell to the ground before Aseneth and pleaded, "Have mercy on us, for you are our mistress and queen. We have sinned against you and against our brother Joseph. Now God has judged us for our wrongdoing.

Please, spare us and save us from our brothers, for they will surely take revenge on us. Their swords will be turned against us, and we will not escape their anger."

Aseneth, moved by their desperation, spoke gently to them. "Do not be afraid. Your brothers serve God, and they do not repay evil with evil.

But for now, go into the woods to keep yourselves safe. What you have done is no small offense, and I must speak to them on your behalf to calm their anger.

Be strong and do not fear, for the Lord Himself will bring justice between us."

Hearing her words, Dan and Gad quickly fled into the woods.

Just then, the sons of Leah came running like swift deer, chasing after them. Aseneth stepped down from her chariot and went to meet them, tears in her eyes.

Her brothers stopped, bowed before her, and wept. Then they asked where the sons of Bilhah and Zilpah had gone, determined to hunt them down and kill them.

Wiping her tears, Aseneth said, "Spare your brothers and do not harm them. The Lord has already defended me. He turned their swords into dust, and they melted away like wax in fire.

That alone should be enough for us. The Lord has fought for us, so there is no need for revenge. Please, let them live."

Simeon, still furious, replied, "Why do you ask us to spare those who have become our enemies?

No! We must kill them with our swords. This is the second time they have plotted evil—first against our father Israel and our brother Joseph, and now against you."

But Aseneth stood firm. "No, my brother," she said. "Do not repay evil with evil. Leave justice in the hands of the Lord. He will punish them in His own time."

Hearing her words, Simeon lowered his head and bowed to her in respect. Then Levi stepped forward, took Aseneth's right hand, kissed it, and blessed her.

Because of Aseneth's wisdom and kindness, she was able to save the sons of Bilhah and Zilpah from their brothers' anger and prevent their deaths.

Chapter Twenty-nine

Pharaoh's son struggled to sit up from the ground. Blood poured from the wound on his temple, running into his mouth as he coughed and gasped for air, spitting out blood.

Benjamin stepped toward him, gripping Pharaoh's sword and pulling it from its sheath, since he had no weapon of his own. With the sword raised, he prepared to strike him down.

Just as Benjamin was about to deliver the final blow, Levi rushed forward and grabbed his hand. "No, my brother," Levi said firmly. "You must not do this. We are men who serve God, and it is not right to repay evil with evil. It is wrong to strike a man who is already defeated or to chase an enemy to his death.

Instead, let's tend to his wound. If he survives, he may become our ally, and his father, Pharaoh, will continue to treat us as his own family."

Levi then helped Pharaoh's son to his feet, gently washed the blood from his face, and wrapped his wound with a bandage. After making sure he was stable, Levi placed him on his horse and brought him back to Pharaoh.

When they arrived, Levi explained everything that had happened. Pharaoh listened carefully, and when he heard the full story, he was deeply moved. Overcome with gratitude and humility, he stood up from his throne and bowed low to Levi.

Despite their efforts to save him, Pharaoh's son died three days later from the injury caused by the stone Benjamin had thrown. Pharaoh was heartbroken over the loss of his eldest son, filled with sorrow and grief.

His sadness took a toll on his health, and he became weaker. Eventually, Pharaoh passed away at the age of 109. Before his death, he placed the crown of Egypt in Joseph's hands.

Joseph ruled Egypt for 48 years, leading the nation with wisdom and fairness. When his time as king came to an end, he passed the crown to Pharaoh's grandson and guided the young ruler, acting as a father to him as he took on the responsibilities of leading Egypt.

Revelation of Elijah

Introduction

Throughout human history, few texts have captivated readers as intensely as apocalyptic literature—writings that unveil hidden truths about the end of the world, judgment, and the ultimate triumph of good over evil. Among these intriguing texts, the Revelation of Elijah holds a unique position, merging Jewish and early Christian apocalyptic traditions into an enthralling vision that vividly portrays cosmic conflict, divine justice, and the fate awaiting humanity in the final days.

The Revelation of Elijah is not merely a prophetic text; it is a compelling narrative that integrates themes of ethical guidance, moral admonition, and visionary insight into a cohesive message. Although traditionally attributed to the prophet Elijah, renowned for his boldness in confronting corruption and idolatry, scholars widely acknowledge that the work originated much later, likely between the 1st and 4th centuries CE. Nevertheless, its association with Elijah—whose fiery spirit and unwavering faith made him a quintessential figure of divine intervention—bestowed upon this text considerable authority and resonance among its ancient readers.

Despite its significance and enduring appeal, the Revelation of Elijah remains one of the lesser-known apocalyptic texts within mainstream biblical scholarship. Its obscurity, however, should not overshadow its profound insights into the early apocalyptic imagination, its rich theological symbolism, and its continuing relevance for contemporary readers seeking understanding amid times of uncertainty. In this introduction, we will delve deeply into the historical background, literary structure, theological themes, and lasting impact of this remarkable text. Our aim is to equip you with essential knowledge and context, ensuring you experience the full depth and significance of Elijah's profound revelations.

Historical Context and Origins

To fully appreciate the significance of the Revelation of Elijah, it is vital to understand the historical environment that gave birth to apocalyptic literature. Apocalyptic texts typically emerged during periods marked by profound crisis, persecution, or societal upheaval. These writings provided readers with reassurance, offering explanations for their suffering and affirming divine justice in the face of chaos. The Revelation of Elijah is no exception, reflecting the turbulent times in which it arose.

Most scholars date the composition of the Revelation of Elijah to between the 2nd and 4th centuries CE, placing it within the complex milieu of late antiquity—a period marked by significant religious tensions, imperial persecutions, and internal conflicts within emerging Christian communities. As early Christians grappled with their identity amidst pagan hostilities and Jewish traditions, apocalyptic texts like this provided clarity, hope, and guidance. Its attributed authorship to Elijah—a prophet revered by both Jewish and Christian communities—further highlights its intent to bridge diverse traditions and communicate profound spiritual truths to a wide audience.

During this era, religious groups frequently faced persecution under Roman authorities, particularly during waves of systematic oppression such as those enacted by Emperors Decius and Diocletian. The apocalyptic genre thus flourished as believers sought explanations for their suffering, assurance of eventual divine intervention, and the promised vindication of the faithful. The Revelation of Elijah echoes these themes powerfully, vividly describing trials, tribulations, and ultimate salvation, encouraging readers to remain steadfast in their faith despite intense hardships.

The setting of persecution and internal strife explains why the text places significant emphasis on ethical behavior, faithfulness, and the rewards awaiting the righteous. These themes not only provided solace but also served as practical instructions for believers navigating their daily lives amidst a hostile environment. By situating readers within a cosmic narrative that transcends immediate struggles, the text reinforced collective identity and resilience, fostering spiritual fortitude essential for community survival and cohesion.

Literary Structure and Key Themes

The Revelation of Elijah is structured around a series of powerful visions and divine dialogues, designed to reveal hidden spiritual truths and outline the progression of eschatological events. Unlike other apocalyptic texts that primarily focus on distant prophecies, the narrative style of this work frequently engages the reader directly, blending vivid imagery, dialogues with angels, and poignant moral teachings.

The text begins by setting the stage for the eschatological drama, introducing Elijah as a prophetic voice chosen by God to deliver revelations about impending judgment and divine intervention. Through symbolic language and evocative imagery, readers are drawn into a gripping account of cosmic warfare between good and evil, with angelic hosts combating demonic forces, and humanity caught in the balance. The imagery is intense, intended to stir both awe and moral urgency in readers, compelling them to consider their own spiritual state in anticipation of the coming judgment.

Among the most notable themes explored in the Revelation of Elijah is the clear delineation between righteousness and wickedness, underscored by ethical exhortations that emphasize purity, compassion, justice, and steadfast faith. Elijah's revelations explicitly outline the moral expectations placed upon believers, warning of the severe consequences awaiting those who persist in injustice, oppression, and idolatry. At the same time, the righteous are promised divine protection, vindication, and reward, creating a powerful incentive for ethical living amidst adversity.

Another crucial aspect of the text is its exploration of divine justice. The work strongly underscores the concept that God's judgment, though delayed, is certain and impartial. The narrative vividly describes punishments reserved for oppressive rulers, false teachers, and all who exploit and corrupt others. Conversely, it portrays the divine deliverance and glorification of the righteous, underscoring the ultimate triumph of God's justice over worldly injustice. By depicting a detailed vision of final judgment and retribution, the text reinforces the importance of perseverance, integrity, and trust in divine providence.

Moreover, the Revelation of Elijah incorporates rich eschatological symbolism, including references to cosmic disturbances, celestial signs, and supernatural events designed to indicate the imminence of the final judgment. Readers are encouraged to interpret contemporary events as meaningful signs of the approaching eschaton, instilling a sense of urgency and vigilance. This eschatological awareness reinforces a shared spiritual identity, aligning believers' personal struggles with broader cosmic events and the overarching divine plan.

Translated by Tim Zengerink

Modern Relevance and Spiritual Insight

Although rooted in ancient traditions and historical contexts, the Revelation of Elijah continues to offer profound insights for modern readers navigating contemporary challenges. Its themes of justice, moral responsibility, perseverance in suffering, and ultimate vindication remain universally relevant, resonating deeply in a world often plagued by injustice, suffering, and uncertainty.

In an era characterized by global unrest, widespread anxiety, and existential questioning, this text speaks powerfully to the human longing for clarity, purpose, and assurance. Its vision of divine justice and final judgment provides comfort, reminding readers that injustice and oppression will ultimately be addressed by a higher power. Its ethical exhortations challenge contemporary readers to consider their roles in promoting compassion, integrity, and social justice in their communities, reinforcing the timeless value of righteous living.

Furthermore, the vivid symbolic imagery and apocalyptic visions within the text serve as powerful metaphors for inner spiritual struggles and personal transformation. The cosmic battles between angelic and demonic forces can be interpreted metaphorically as struggles against personal vices, societal injustices, and spiritual darkness. Thus, the text not only describes external, historical events but also encourages introspection, personal reflection, and inner spiritual growth.

Finally, the Revelation of Elijah encourages vigilance and preparedness, emphasizing the unpredictability and suddenness of divine judgment. In a contemporary context, this translates into a heightened awareness of the impermanence and fragility of human existence, fostering a deeper appreciation of life's inherent value and the urgency of ethical decision-making.

In summary, the Revelation of Elijah stands as a powerful testament to humanity's enduring quest for spiritual truth, ethical guidance, and hope amidst adversity. As you embark upon your journey through this text, may its vivid revelations inspire reflection, deepen understanding, and strengthen your resolve to live with integrity, compassion, and unwavering faith in the ultimate triumph of divine justice.

Revelation of Elijah

I

The Lord spoke to me and said, "Son of man, tell the people: Why do you keep piling up your sins and making your Creator even angrier?"

He warned us not to love the world or chase after the things in it. The pride and downfall of the world come from the devil.

We should remember that the powerful Lord, who made everything, showed us mercy to free us from the grip of this age.

The devil has often tried to stop the sun from rising and the earth from growing food because he wants to destroy people.

He's like a wildfire or a flood that wipes everything out.

But God, full of mercy, sent His Son to rescue us.

He didn't send an angel or heavenly power—He came Himself, becoming human, so He could save us.

So now be His children, because He's your Father.

God has prepared thrones and crowns in heaven for those who obey Him. He said, "Whoever follows Me will sit with My people and wear a crown."

He also said, "I will write My name on their foreheads and put a mark on their right hands. They won't be hungry or thirsty anymore.

Evil won't have power over them. They'll walk with angels into My holy city."

But those who live in sin will be full of shame. They won't make it past the thrones.

Death will grab hold of them, and the angels won't stand with them.

They've turned their backs on the home of God.

Listen carefully, wise people, about those who will appear near the end of time.

These deceivers will spread teachings that aren't from God.

They'll ignore God's Law and follow their own desires.

They'll say things like, "Fasting isn't real," and deny that God created it.

By doing this, they cut themselves off from God's promises and His agreement with us.

They never truly had a strong faith, so don't let them mislead you.

From the beginning, when God made the heavens, He gave us fasting as a gift to help fight our temptations and desires.

God said, "I created true fasting."

Someone who fasts with a clean heart won't fall into sin, even when dealing with jealousy or conflict.

Let the person who is pure fast. But if someone fasts with an impure heart, it makes God and His angels angry.

He only hurts himself and piles up punishment for judgment day.

Real fasting comes from a clean heart and clean hands.

It erases sin, heals sickness, and pushes out demons.

It rises up to God like a healing prayer that brings forgiveness.

Think about it: who would go to work without their tools? Or go into battle without armor?

Wouldn't they be punished for not following the king's orders?

In the same way, you can't stand before God if your heart is divided.

A person like that walks in darkness, and even the angels don't trust them.

So stay focused on the Lord all the time so you'll understand what's happening and what to do.

II

Now let's talk about the kings of Assyria and the final destruction of the heavens, the earth, and the world below.

The Lord says, "My people won't be defeated or scared in battle."

You'll see a king rise in the north—he will be called the king of Assyria and the king of injustice.

He'll cause chaos and war in Egypt.

The land will suffer, and your children will be taken away.

Many people will wish for death, but it won't come.

Another king, known as the king of peace, will rise from the west.

He'll cross the sea like a roaring lion.

He'll kill the king of injustice and bring bloody revenge to Egypt.

Later, he'll talk about peace and offer gifts in Egypt.

He'll speak kindly to God's people and say, "There is only one true God."

He'll honor the holy ones and lift up their sacred places.

But the gifts he offers to God's house will be empty and false.

He'll trick people in Egypt without them knowing it.

He'll study the holy places and the idols and count all their treasure.

He'll assign new priests and arrest the wise and powerful people, sending them to a large city by the sea, saying, "Now we all speak one language."

But when people start saying, "We have peace and joy," be alert…

Here's how you'll know who this king is:

He has two sons, one on each side of him.

The son on his right will have a wicked face and will fight against God.

Four more kings will come from this line.

In his thirtieth year, the king will go to Memphis and build a temple.

But that same day, his own son will kill him.

This will throw the whole land into confusion.

He will order that all priests and holy people be captured.

He'll demand they repay double for everything his father gave them.

He'll close the holy places, steal their homes, and take their sons away.

He'll command evil sacrifices to be made in the land.

He will appear in the sky in front of the sun and moon.

That day, the priests will rip their clothes in sorrow.

Woe to you, rulers of Egypt. Your time is over.

The harm you did to the poor will fall back on you, and your children will be taken.

Egypt's cities will cry out. Buying and selling will stop.

Markets will be empty and silent.

People will cry for death, but death will escape them.

They'll run to cliffs and jump, shouting, "Fall on us!"—but they still won't die.

The whole land will suffer twice as much.

The king will command that all women who are nursing babies be captured.

They'll be forced to feed snakes, and their blood will be turned into poison for arrows.

Because of the suffering in the city, he'll take all boys twelve and under and train them for war.

Midwives will mourn. Mothers will lift their eyes to heaven and say, "Why did I give birth to a son?"

Women who never had children will be happy, saying, "Now is our time to rejoice—our children are safe in heaven."

During this time, three kings from Persia will rise and bring back the Jews living in Egypt to Jerusalem. They will settle there again.

When you hear that there's peace in Jerusalem, priests of the land, tear your robes—because the destroyer is near.

At that time, the evil one will enter the holy places—

Then Persian kings will quickly come to fight against the kings of Assyria. Four kings will fight three.

The war will last for three years until they steal the treasures of the temple.

Blood will flow from Kos to Memphis. The river in Egypt will turn to blood, and no one will be able to drink it for three days.

Woe to Egypt and everyone in it.

A new king will rise in the "City of the Sun," and the land will fall into chaos. He'll escape to Memphis with the Persians.

In the sixth year, Persian kings will ambush and kill the Assyrian king in Memphis.

The Persians will take revenge. They'll kill the wicked, rebuild the holy places,

and give double offerings to God's house, saying, "The Lord is one."

Everyone will praise the Persians.

Those who survive will say, "The Lord gave us a good king so our land wouldn't be destroyed."

That king will declare that no royal orders will be given for three and a half years.

The land will be full of peace and plenty.

People who are still alive will call out to the dead, saying, "Come join us in this peaceful rest."

III

In the fourth year of that king's reign, an evil man will appear and say, "I am the Christ." But don't believe him—he's not telling the truth.

When the true Christ comes, it will be gentle and peaceful, like a group of doves flying with Him. A cross will go before Him as He moves across the sky.

Everyone in the world will see Him, just like we all see the sun rise in the east and set in the west.

That's how He'll return—surrounded by angels.

But the evil man will try to stand in God's holy places again.

He'll show off, saying things like, "Sun, fall!"—and the sun will fall.

He'll say, "Shine!"—and it will shine.

He'll say, "Go dark!"—and the sun will go dark.

He'll tell the moon to turn red like blood, and it will happen.

He'll come down from the sky and walk on seas and rivers like they're solid ground.

He'll make people who can't walk start walking.

He'll make the deaf able to hear.

He'll help people who can't speak to talk.

He'll help the blind to see.

He'll heal skin diseases and make the sick well.

He'll cast out demons too.

He'll do many miracles that people can see with their own eyes.

He'll copy a lot of what Jesus did—except he won't be able to raise the dead.

That's the way you'll know he's not the real Christ—because only God can give life.

Here's what he will look like, so you can recognize him:

He'll be a skinny young man with weak-looking legs.

He'll have a patch of gray hair at the front of his bald head.

His eyebrows will stretch all the way to his ears.

He'll have a bare, sick-looking spot on the front of his hands.

He'll be able to change how he looks in front of people.

Sometimes he'll look like a child. Other times, he'll look like an old man.

He'll try to change everything about his appearance—but his head will stay the same.

That's how you'll know he's the evil one.

IV

A young woman named Tabitha will hear that the evil one has shown up in God's holy places. She'll dress in clean, fine clothes and chase after him all the way to Judea, shouting at him all the way to Jerusalem: "You shameful liar! You enemy of all God's people!"

This will make the evil one furious. He'll chase her west, all the way to where the sun sets. That evening, he'll hurt her and spill her blood. Then he'll throw her body onto the temple grounds—but her blood will become a healing for the people.

At sunrise, she'll come back to life. She'll stand up and shout at him again, saying, "You may have attacked me, but you have no power over my soul or my body. I live in the Lord always. Even the blood you spilled became a blessing for others."

When Elijah and Enoch hear that the evil one is standing in God's holy place, they'll come down from heaven to face him. They'll speak directly to him:

"Aren't you ashamed? You try to join God's people, even though you never belonged to them. You've always been against heaven and against those on earth too. You've attacked the thrones and gone against the angels. You're nothing but a stranger.

You were thrown out of heaven like a falling star. You lost your light, and your followers are now in darkness. And even now, you're not ashamed to stand against God. You are the devil."

The evil one will hear them and get furious. He'll start a fight with them in the main square of the great city, and the battle will last for seven days.

Elijah and Enoch will die, and for three and a half days, their bodies will lie in the marketplace where everyone can see them.

But on the fourth day, they'll rise again. They'll confront him, saying, "You shameful liar! Aren't you embarrassed? You're leading God's people astray, and you never even suffered for them. Don't you realize we live in the Lord?"

Then they'll say, "We'd rather give up our lives for the truth than be like you. We will defeat you because you have no real strength. We are strong in the Lord, and you are always against Him."

The evil one will become even more enraged and attack them again.

The whole city will gather around them. But Elijah and Enoch will shine like the sun and call out to heaven. Everyone in the city—and in the whole world—will see them.

The son of lawlessness won't be able to defeat them. But he'll turn his anger toward the land and the people.

He'll chase after all the saints. He'll capture them, along with the priests, and tie them up.

Then he'll kill and torture them. He'll gouge out their eyes with iron spikes. He'll rip the skin from their heads and pull out their fingernails one by one. He'll order that vinegar and lime be poured into their noses.

Some of the people who can't take the pain will run to the desert with whatever gold they can carry. There, they'll lie down as if asleep.

The Lord will take their spirits and souls to Himself. Their bodies will turn to stone, and wild animals won't touch them until the final day of judgment.

On that day, they'll rise and find peace—but they won't receive the same reward as those who stayed strong through the suffering. The Lord said, "I will let those who stayed faithful sit at My right hand."

These strong ones will be honored above the others. They'll win the fight against the son of lawlessness. They will see heaven and earth pass away, and they will be given glorious thrones and shining crowns.

At that time, sixty righteous people—already chosen for this moment—will hear the call.

They'll put on the armor of the Lord and rush to Jerusalem to face the evil one. They'll say, "You've done all the miracles the prophets did in the past. But you couldn't raise the dead, because you don't have the power to give life. That's how we know you're the son of lawlessness."

He'll hear this, get furious, and order altars to be lit on fire.

The righteous ones will be captured, tied up, lifted high, and burned.

V

On that day, many people will open their eyes and run from the evil one. They'll say, "This isn't the real Christ. The true Christ doesn't kill good people or hunt them down. He brings people to Him through love and miracles, not by force."

That day, the real Christ will show kindness to His people. He will send 64,000 angels from heaven, each with six wings. Their voices will shake the sky and the earth as they praise God.

Everyone who has Christ's name written on their forehead and His mark on their hand—young or old—will be lifted up by the angels and taken away before the anger of God comes.

Gabriel and Uriel will appear as shining lights and lead them into the holy land.

There, they'll eat from the tree of life and wear white clothes. Angels will guard them. They won't thirst, and the evil one won't be able to harm them.

That day, the earth will be thrown into chaos. The sun will go dark. Peace will vanish. Birds will drop from the sky. The land will dry up. Even the sea will lose its water.

The sinners will cry out, "What have you done to us? You said you were Christ, but you're not—you're the devil! You couldn't even save yourself, so how could you save us?

You fooled us with your tricks and pulled us away from the real Christ who made us. Now we're doomed because we believed you.

We're starving. Who can we trust now? Who will lead us or help us? No one.

We're being punished because we turned away from God. We searched the sea and the rivers, but couldn't find water."

Then the evil one will cry out too. He'll say, "No! My time is over. I kept saying I'd rule forever, but it's too late. My years became months, and my days disappeared like dust. Now I'll be destroyed along with you."

He'll yell, "Go to the desert! Catch the thieves and kill them! Bring back the saints! They're the reason the earth still grows food. Because of them, the sun still shines and the dew still falls."

But the sinners will weep and say, "You turned us against God! If you're so powerful, why don't you chase them yourself?"

Then the evil one will spread his fiery wings and fly after the saints. He'll fight them again.

The angels in heaven will hear and come down to fight back. There will be a huge battle with swords.

That day, the Lord will speak, and both heaven and earth will answer with anger. Fire will be sent down.

It will burn across the land for 72 cubits, destroying the wicked and demons like dry grass in flames.

This will be God's final judgment.

Even the mountains and roads will speak, saying, "Did you hear someone walk past us today who didn't face God's judgment?"

Every sin will show itself—whether done in the day or night—and it will appear right where it happened.

The righteous, along with those who were hurt or killed for their faith, will see their enemies suffer.

Even the sinners being punished will see the place where the righteous live.

God will be generous. In those days, He will give the righteous everything they've asked for many times before.

Then the Lord will judge the heavens and the earth. He'll bring justice to everyone who sinned—whether above or below.

He'll also judge the leaders, asking what they did for the people they were supposed to care for. The people who were truly faithful will be given back to God, clean and free from evil.

After this, Elijah and Enoch will return. They'll leave behind their old bodies and receive new, heavenly ones.

They'll hunt down the evil one and defeat him, and he won't be able to speak anymore.

Right in front of them, he'll melt like ice near a fire. He'll disappear like a lifeless snake.

They'll say to him, "Your time is done. Now you and everyone who followed you will be wiped out."

Then he and his followers will be thrown into a deep pit, and it will be locked forever.

That same day, Christ the King and all His saints will come down from heaven.

He will burn away the old world and spend a thousand years on earth.

Because evil people ruled over it, He will make a brand new heaven and a new earth where no sin will ever exist again.

He will rule with His saints, going back and forth from heaven to earth. They will be with the angels and with Christ for one thousand years.

Revelation of Zephaniah

Introduction

Apocalyptic literature has long fascinated readers, theologians, and scholars alike, as it explores some of the deepest and most compelling questions about human existence: the nature of good and evil, the ultimate fate of humanity, the role of divine judgment, and the mysteries of life beyond this world. Amidst the vast and varied tapestry of ancient texts, few are as intriguing and enigmatic as the Apocalypse of Zephaniah. Though less familiar to contemporary readers than some canonical apocalyptic texts, this remarkable work provides a unique window into the complex interplay of early Jewish and Christian eschatological thought. It offers powerful, vivid imagery, profound theological reflections, and a gripping narrative structure designed not only to inform but also to transform readers through its visionary experience.

Attributed traditionally to Zephaniah, a minor prophet recognized within the Hebrew Bible, this apocalypse is rooted in the ancient tradition of prophetic revelation. Despite this attribution, scholarly consensus places its composition considerably later than the historical Zephaniah, situating it firmly in the tumultuous religious and cultural milieu between the 1st century BCE and the 2nd century CE. During this period, Jewish communities were profoundly impacted by Roman oppression, internal divisions, and apocalyptic expectations intensified by the struggles for identity and survival. In this historical crucible, the Apocalypse of Zephaniah emerged as a powerful response, addressing the fears, hopes, and spiritual aspirations of its original audience.

The text itself is deeply visionary, weaving together celestial journeys, angelic guides, heavenly revelations, and vivid descriptions of judgment, reward, punishment, and redemption. Zephaniah, guided by divine messengers, journeys through realms previously hidden from human eyes. Readers are transported alongside the prophet through dramatic revelations of the afterlife, divine justice, and cosmic mysteries. As such, the Apocalypse of Zephaniah stands as a vital bridge linking earlier prophetic traditions with later Christian and Jewish mystical thought, contributing significantly to the tapestry of apocalyptic literature that continues to influence spiritual and cultural imagination today.

In this comprehensive introduction, we aim to equip readers with essential historical context, a deep understanding of the literary themes and symbolic structures within the text, and a clear appreciation of its enduring relevance. By the conclusion, readers will not only have gained critical insights into the Apocalypse of Zephaniah but will also be better prepared to engage deeply with its powerful and transformative narrative.

Historical and Cultural Background

To grasp the full significance of the Apocalypse of Zephaniah, one must first appreciate the historical and cultural circumstances that gave rise to its creation. Apocalyptic writings traditionally arise during times of great upheaval, persecution, and crisis. They aim to reassure communities under duress, providing theological explanations for suffering and affirmations of eventual divine justice and redemption. The Apocalypse of Zephaniah emerged precisely within such a context, reflecting profound social, religious, and political challenges faced by early Jewish and emerging Christian communities in the late Second Temple and early post-Temple periods.

Most scholars date the composition of this text to roughly between the late 1st century BCE and the early 2nd century CE. This era was particularly tumultuous: Jewish communities had recently endured the trauma of the Roman destruction of the Second Temple in 70 CE, leading to profound existential and theological crises. The catastrophic loss of the Temple—central to Jewish identity, ritual practice, and the perceived presence of God—caused widespread spiritual despair and deep theological questioning. Additionally, Jewish communities faced increasing tensions both internally, through sectarian divisions, and externally, from intensifying Roman persecution.

Within this historical framework, apocalyptic literature served as a crucial mechanism for both spiritual consolation and theological guidance. Texts such as the Apocalypse of Zephaniah provided readers with reassurance that, despite present suffering, divine justice was assured, and the righteous would ultimately experience vindication and reward. Its visionary narratives offered meaning and hope amidst despair, articulating a powerful theological response to crisis and affirming the continuity of divine purpose in human history.

Furthermore, the historical Zephaniah, whose prophetic ministry occurred during the reign of King Josiah of Judah (approximately 640–609 BCE), had already established a tradition of vividly describing divine judgment, redemption, and restoration. The attribution of this apocalypse to Zephaniah provided significant authority and credibility, ensuring that its messages resonated deeply with readers familiar with his canonical prophecies. By leveraging Zephaniah's prophetic reputation, this text effectively bridged earlier Hebrew prophetic traditions with emerging eschatological expectations, blending ancient authority with contemporary relevance.

In summary, the historical and cultural backdrop of the Apocalypse of Zephaniah is essential for understanding its enduring power. Its themes of judgment, justice, divine sovereignty, and ultimate redemption were shaped profoundly by the existential crises and cultural struggles of the late Second Temple and post-Temple era, making it a vital testament to the resilience of faith and hope amidst profound adversity.

Literary Structure and Major Themes

The literary structure of the Apocalypse of Zephaniah is distinctive, effectively combining narrative visions, angelic dialogues, symbolic revelations, and detailed accounts of celestial journeys. Such visionary apocalyptic literature often engages readers through a journey format, inviting them to accompany the prophet through heavenly realms, supernatural events, and eschatological disclosures. This structural approach heightens both dramatic tension and emotional engagement, as readers are drawn progressively deeper into increasingly profound revelations.

The text begins by introducing Zephaniah as a divinely appointed seer who is granted extraordinary access to heavenly mysteries and eschatological truths. Guided by angelic beings, Zephaniah traverses unseen realms, encountering astonishing scenes of heavenly judgment, spiritual trials, and cosmic phenomena. This journey structure not only reinforces the authority of the prophet but also creates an intimate bond between Zephaniah and the reader, who becomes a fellow traveler in the prophet's visionary experiences.

Central to the thematic framework of the Apocalypse of Zephaniah is the portrayal of divine judgment and justice. The text vividly details both the punishments awaiting the wicked and the rewards reserved for the righteous. Its descriptions of the afterlife, heavenly courts, and divine retribution are rich with symbolism and metaphor, designed to convey moral clarity and urgency. Readers are continually reminded of the absolute impartiality and inevitability of divine judgment, underscoring a powerful call to ethical accountability and moral vigilance.

A particularly compelling theme within the text is the portrayal of the soul's journey after death, making it one of the earliest and most detailed ancient Jewish accounts of postmortem experiences. The Apocalypse of Zephaniah provides vivid descriptions of the judgment of souls, depicting angelic beings administering justice and illustrating the fates awaiting individuals based on their earthly lives. These descriptions profoundly influenced later Jewish and Christian conceptions of heaven, hell, and divine justice, enriching theological discourse about life after death.

Moreover, the text is notable for its sophisticated angelology, which details hierarchies and roles of angels within the divine economy. The angels depicted are not merely messengers but active participants in cosmic justice, guardians of moral order, and executors of divine decrees. By incorporating this complex angelology, the text emphasizes the cosmic scale of spiritual struggles, reinforcing the belief that earthly realities reflect deeper spiritual conflicts.

Enduring Significance and Contemporary Relevance

Though composed in ancient times, the Apocalypse of Zephaniah maintains enduring relevance and resonance for contemporary readers. Its profound themes of judgment, justice, moral responsibility, and spiritual destiny speak powerfully to modern audiences grappling with similar existential questions and ethical dilemmas.

In an age marked by widespread social injustice, moral ambiguity, and existential uncertainty, the text's affirmation of ultimate divine justice provides a compelling moral compass. Its vivid portrayals of ethical consequences remind contemporary readers that individual actions carry profound spiritual significance, underscoring the importance of ethical accountability and moral integrity. Its descriptions of cosmic justice also inspire hope, reassuring readers that the seemingly relentless injustices of the present age will ultimately be rectified.

Furthermore, the text's depiction of afterlife journeys resonates deeply with contemporary spiritual curiosity about life beyond death. Its vivid imagery and narrative power serve as metaphors for personal spiritual transformation, encouraging introspection and moral reflection. By engaging readers in its profound spiritual insights, the Apocalypse of Zephaniah invites a deeper exploration of life's ultimate meaning and purpose.

Ultimately, the Apocalypse of Zephaniah stands as a remarkable testament to humanity's enduring quest for spiritual understanding, moral clarity, and existential hope. As readers embark upon this ancient yet profoundly relevant visionary journey, they will find themselves inspired, challenged, and transformed by the timeless wisdom and powerful revelations of this extraordinary text.

Revelation of Zephaniah

Clement, Stromata 5.11.77

What I saw in the fifth heaven.

A spirit carried me up into the fifth heaven. There, I saw angels called "lords." Each one wore a glowing crown given by the Holy

Spirit. Their thrones shone with a light that was seven times brighter than the sunrise. They lived in holy places where people are saved, and they sang songs of praise to God, whose greatness is beyond words.

Sahidic fragment

A soul punished in a vision.

I saw a soul being punished and guarded by five thousand angels. They dragged it from the East to the West, beating it again and again—hundreds of times each day. I was so frightened that I fell to the ground, and it felt like my body was falling apart.

An angel came and helped me up. He said, "Be strong. You are someone who will overcome. You will defeat the one who accuses you and rise from the place of the dead."

After I got up, I asked him, "Who is this soul they're punishing?"

He answered, "This soul lived in sin. Before it could repent, it was taken from its body."

I, Zephaniah, truly saw all of this in my vision.

What I saw in a wide open space.

The angel of the Lord stayed with me as I continued. I came to a huge open area. On the left side, there were thousands and thousands of beings. On the right side, there were even more—too many to count. Each one looked different from the others. They all had long, flowing hair like women. Their teeth were sharp and fierce, unlike anything I'd seen before.

Akhmimic Text

A piece about someone being buried.

He is dead now. We'll bury him like we would any other person.

When the time comes, we'll carry him out, playing music on the harp and singing psalms and songs over his body.

What was seen above the seer's city.

I went with the angel of the Lord, and he lifted me up high above my city. I couldn't see anything around me at first.

Then I saw two men walking together on a road, talking as they went. I also saw two women grinding grain at a mill, chatting with each other. Then I saw two people lying on a bed, each one focused on the other.

After that, I looked and saw the whole world. It was hanging in the air like a single drop of water dangling from a bucket being pulled up from a well.

I asked the angel, "Is there no night or darkness here?"

He answered, "No. There's no darkness in the place where the good and holy people live. There, it's always filled with light."

Then I saw all the souls of people who were being punished. I cried out to God, "Lord, if You stay close to the saints, then surely You also care about the world and all the souls who are suffering right now."

Angels keeping records on Mount Seir

The angel of the Lord said to me, "Come, I'll show you the place where the righteous live." He brought me up to Mount Seir, where I saw three men. Two angels were walking with them, celebrating and full of joy.

I asked the angel, "Who are these people?"

He answered, "These are the three sons of Joatham the priest. They didn't follow their father's teachings or obey God's commands."

Then I saw two other angels crying over those three sons.

I asked, "Who are these angels?"

He said, "They are God's angels. They write down all the good things that righteous people do. They stand at the gate of heaven and keep careful records. I take what they've written and bring it to the Lord, and He writes their names in the Book of Life."

He continued, "But there are also angels working for the one who accuses people. They write down all the wrong things people do and also wait at the gate of heaven. They report everything to the accuser, who writes it on his scroll so he can speak against them when their souls leave this world and go below."

Frightening angels take away the souls of wicked people.

I was walking with the angel of the Lord when I looked ahead and saw a place. Huge crowds of angels were going into it—too many to count.

Their faces looked like leopards, and they had tusks like wild boars sticking out of their mouths. Their eyes were red and bloodshot. Their long hair flowed like a woman's, and they held burning whips in their hands.

When I saw them, I got really scared. I asked the angel walking with me, "Who are these beings?"

He answered, "These are the ones sent to deal with the souls of wicked people. They take those souls and bring them here. For three days, they carry them through the air, then throw them into their eternal punishment."

I said, "Please, Lord, don't let them come near me!"

The angel said, "Don't be afraid. I won't let them touch you. You are clean in the eyes of the Lord. He sent me to be with you because you are pure before Him."

Then the angel waved at them, and they backed away and ran off.

The city in heaven.

I walked with the angel of the Lord and saw some gates ahead of us. As we got closer, I realized the gates were made of bronze. The angel touched them, and they opened right away. I went in with him and saw that the whole area looked like a beautiful city. I walked through the middle of it.

Then, right beside me, the angel of the Lord changed form.

I looked around and saw more bronze gates, locked with bronze bolts and iron bars. I was so amazed I couldn't speak. I stood there, staring at the gates as fire came out from them and stretched about fifty stadia into the distance.

The accuser and the angel Eremiel in the place of the dead.

I turned around and started walking again. I saw a huge sea in front of me. At first, I thought it was made of water, but I realized it was actually a sea of fire—thick and bubbling like slime, with flames shooting out and waves burning with sulfur and tar. The fiery sea started moving toward me.

I thought God Himself was coming to meet me. I dropped to the ground to worship, completely terrified. I begged Him to rescue me from the fear I was feeling. I cried out, "God, Lord, please help me. Save me from this awful danger."

Right after that, I stood up and saw a huge angel standing in front of me. His hair was wild, like a lion's mane. His teeth stuck out like a bear's. His long hair flowed like a woman's, and his body looked like a snake ready to swallow me.

I was so scared that I collapsed. My whole body felt weak, and I couldn't stand up. I prayed, "Lord, save me from this. You saved Israel from Pharaoh in Egypt. You helped Susanna when the evil elders tried to hurt her. You saved Shadrach, Meshach, and Abednego from the burning fire. Please save me too."

Then I managed to get up again. I saw another great angel standing there. His face was bright like the sun, shining with glory. He wore something like a golden sash across his chest, and his feet looked like glowing bronze, hot from the fire.

When I saw him, I felt so happy. I thought God had finally come to visit me. I fell on my face to worship him.

But he said, "Wait. Don't worship me. I'm not the Lord. I'm the great angel Eremiel. I watch over the deep pit—Hades—where the souls are kept, from the time of the great flood until now."

Then I asked him, "Where am I?"

He answered, "You're in Hades."

I asked again, "Who was that other great angel I saw?"

He said, "That's the one who accuses people in front of God."

The two scrolls.

I saw someone holding a scroll and watched as he slowly unrolled it.

When it was open, I realized I could read everything—it was in my own language. Every mistake I had made, from when I was a kid up to that moment, was written on it. Nothing was left out, and nothing was untrue.

If I didn't visit someone who was sick or a widow, it was written down as something I had failed to do.

If I ignored an orphan, that was listed too.

If I skipped a day of prayer or didn't fast when I should have, it showed up on the scroll.

Even if I didn't help or care for my own people, that was recorded as well.

When I saw all of it, I dropped to the ground and begged God, "Please, show me kindness. Wipe this scroll clean. I know Your mercy is everywhere and reaches everything."

Then I got up, and a mighty angel stood in front of me. He said, "You've made it through. You defeated the one who tried to accuse you. You've been brought out of the world of the dead. Now you're ready to cross into new life."

Then he brought out another scroll, this one written by hand. He began to open it, and again, I could read it in my own language...

Leaving the place of the dead.

Two pages missing (In the missing pages, the writer likely described what was in the second scroll, which probably listed the good things the seer had done. If it followed the same pattern as the part before, the scroll would have been read, then a prayer—maybe a prayer of thanks—would be offered, and a great angel would announce victory. After that, they would begin getting ready to cross the river on their way out of Hades. The next part starts when the boat arrives.)

They helped me and placed me on the boat.

Thousands upon thousands—more than anyone could count—of angels praised God in front of me.

I put on clothes like the angels wear. I saw all the angels praying.

I joined in and prayed with them.

I understood their language, and they spoke to me in it.

Now, my sons, this is the moment of judgment, because good and evil must be weighed on the scale.

The first trumpet: victory and a visit for the faithful

Then a great angel appeared, holding a golden trumpet. He blew it three times above my head and said, "Be brave, you who have won! Stand strong, you who have overcome! You've defeated the one who accused you, and you've escaped from the deep pit and the place of the dead. Now you will cross over to the other side. Your name is written in the Book of Life."

I wanted to hug him, but I couldn't because his glory was too powerful.

Then the angel went quickly to all the faithful ones—Abraham, Isaac, Jacob, Enoch, Elijah, and David. He spoke with them like a friend talks with another friend.

The second trumpet: heaven opens and the suffering of souls is revealed

Then the great angel came to me with a golden trumpet and blew it up toward the sky. Heaven opened wide—from east to west, and from north to south.

I saw the same sea I had seen before at the bottom of Hades. Its waves reached all the way up to the clouds. I saw many souls sinking into it. Some had their hands tied to their necks, and their arms and legs were chained.

I asked, "Who are these people?"

He said, "These are the ones who took bribes. They accepted gold and silver to mislead others."

Then I saw others covered in mats of fire.

I asked, "Who are these?"

He said, "They are the ones who loaned money and charged unfair amounts of interest—interest on top of interest."

I also saw blind people crying out, and I was overwhelmed by everything I saw.

I asked, "Who are these?"

He said, "They heard God's word, but they didn't live by it or follow through."

I asked, "Can they still repent?"

He answered, "Yes."

I asked, "For how long?"

He said, "Until the day the Lord gives His final judgment."

Then I saw others with hair on their bodies.

I asked, "Do they still have hair and bodies in this place?"

He said, "Yes, the Lord gives them bodies and hair however He wants."

The saints ask for mercy for people who are suffering.

I saw a huge crowd, and the angel brought them closer.

When they looked at all the pain and suffering, they started to pray to the Lord, saying, "Please, have mercy on everyone who's going through this."

I asked the angel with me, "Who are these people?"

He answered, "They are Abraham, Isaac, and Jacob.

Every day, at the same time, they come with a great angel. One angel blows a trumpet toward heaven, and another blows one toward the earth.

All the good and faithful people hear the sound and hurry over. They pray to the Lord every day for those who are trapped in this suffering."

Another trumpet blast warns that God's anger is coming.

The great angel came out again, holding the golden trumpet, and blew it over the earth.

People heard it from the east to the west, and from the south to the north.

Then he blew the trumpet toward heaven, and its sound echoed above.

I said, "Lord, why didn't You let me stay until I saw everything?"

He answered, "I don't have the power to show you the rest until the Lord Almighty rises in anger to destroy the earth and the sky.

When that happens, everyone will see it and be terrified. They will cry out, 'We give You all people that belong to You, Lord, on Your day.'

Who will be able to stand before Him when He rises in anger to bring destruction?

Every tree growing on the earth will be torn up by the roots and fall. Every tall tower and all flying birds will fall too."

Four pages missing

Thank You for Reading

Dear Reader,

We hope this timeless classic has sparked your imagination and enriched your literary journey. Now that you've turned the final page, we want to share a vision for the future of reading—one where every classic you've ever wanted to explore is at your fingertips, in a format that best suits your life.

We'd like to invite you to gain immediate, unlimited digital & audiobook access to hundreds of the most treasured literary classics ever written—along with the option to secure deluxe paperback, hardcover & box set editions at printing cost. Together, we can spark a new global literary renaissance alongside our small, independent publishing house called "The Library of Alexandria."

Thousands of years ago, the Library of Alexandria stood as a beacon of knowledge—until it was lost to history. We aim to reignite that spirit of preservation and discovery right now, in the modern age—only this time, it's accessible to all, in every language and every format.

Picture a world where every timeless classic, novel, poem, or philosophical treatise is not only available to read but also updated for today's readers—modernized, translated into any language or dialect, and ready to enjoy in any format you choose, whether that is in an eBook, audiobook, paperback, or deluxe hardcover & box set version a printing cost.

By joining our movement to rebuild the modern Library of Alexandria, you become part of an unprecedented mission to offer:

- **Unlimited Audiobook & eBook Access to the Greatest Classics of All Time**

  Instantly explore thousands of legendary works, from Plato and Shakespeare to Jane Austen and Leo Tolstoy. All are instantly ready to read or listen to, giving you a complete literary universe at your fingertips.

- **Paperback & Deluxe Editions at Printing Costs:**

  Purchase any title in a paperback, deluxe hardbound, or deluxe boxset edition at printing costs, shipped right to your doorstep. Curate your personal library of Alexandria with editions worthy of display—crafted to last, designed to captivate, and delivered straight to your door.

- **Modern translations for Contemporary Readers in all languages and dialects**

  Discover a vast selection of classics reimagined in clear, current language—no more struggling with outdated phrases or obscure references. Next to the original versions, we aim to offer translations in as many languages and dialects as possible.

  As we continue our translation efforts and add new languages, readers everywhere can connect with these works as if they were written today. By bridging linguistic divides, you're contributing to ensuring that these timeless stories become more meaningful, accessible, and inspiring for people across the globe.

- **Your Personal Library of Alexandria:**

  Over the months and years, you'll curate a unique physical archive of classics—each volume a testament to your taste, curiosity, and love of knowledge. It's not just about owning books—it's about curating a cultural legacy you'll cherish and pass down for generations to come.

- **Join a Global Literary Renaissance:**

    Your support fuels an ongoing mission: allowing us to reinvest in offering deluxe print editions (including special boxsets) at their true cost, broaden the range of available formats and translations, and extend the reach of these works to new audiences worldwide. By joining today, you're not just preserving a legacy of masterpieces; you set in motion a powerful wave of literary accessibility.

    We are more than a publisher—we're a movement, and we can't do it alone. Your support lets us scale our mission, preserving and reimagining history's greatest works for tomorrow's readers.

Become a Torchbearer of knowledge.

Thank you for picking up this book and allowing us into your literary journey. As you turn the pages, know that you're part of something larger: a global effort to keep these stories alive, share their wisdom across borders and generations, and spark a true cultural revival for the modern era.

If this resonates with you—please consider taking the next step by visiting:

www.libraryofalexandria.com

With gratitude and a shared love of knowledge,

The Modern Library of Alexandria Team

Visit:

www.libraryofalexandria.com

Or scan the code below: